David Thomas Ansted

In search of minerals

David Thomas Ansted

In search of minerals

ISBN/EAN: 9783337149642

Printed in Europe, USA, Canada, Australia, Japan

Cover: Foto ©berggeist007 / pixelio.de

More available books at **www.hansebooks.com**

SUNDERLAND

Education Authority.

Valley Road Boys' School.

Presented to

[signature]

FOR REGULAR ATTENDANCE.

H. BROWN, B.A.,
HEAD MASTER.

190...1

Hills & Company, Printers, Sunderland.

NATURAL HISTORY RAMBLES.

IN SEARCH OF MINERALS.

BY

D. T. ANSTED, M.A. F.R.S.

FORMERLY FELLOW OF JESUS COLLEGE, CAMBRIDGE,
HONORARY FELLOW AND LATE PROFESSOR OF GEOLOGY IN KING'S
COLLEGE, LONDON.

PUBLISHED UNDER THE DIRECTION OF
THE COMMITTEE OF GENERAL LITERATURE AND EDUCATION
APPOINTED BY THE SOCIETY FOR PROMOTING
CHRISTIAN KNOWLEDGE.

LONDON:
SOCIETY FOR PROMOTING CHRISTIAN KNOWLEDGE,
NORTHUMBERLAND AVENUE, CHARING CROSS, W.C.
43, QUEEN VICTORIA STREET, E.C.
BRIGHTON: 129, NORTH STREET.
NEW YORK: E. & J. B. YOUNG & CO.
1895.

RICHARD CLAY & SONS, LIMITED,
LONDON & BUNGAY.

CONTENTS.

CHAPTER I.
MINERALS AND THEIR HOMES... ... *page* 1

CHAPTER II.
GEMS OR PRECIOUS STONES — THE DIAMOND AND RUBY 35

CHAPTER III.
PRECIOUS STONES, CONTINUED—THE EMERALD, ZIRCON OR HYACINTH, TOPAZ, SPINELLE, TOURMALINE, AND GARNET GROUPS ... 68

CHAPTER IV.
QUARTZ GEMS 92

CHAPTER V.
THE SOFTER GEMS AND VALUABLE STONES ... 124

CHAPTER VI.

VALUABLE MINERALS DERIVED FROM THE VE-
GETABLE AND ANIMAL KINGDOMS;—JET,
AMBER, PEARL, CORAL ... ... *page* 156

CHAPTER VII.

USEFUL NON-METALLIC MINERALS 185

CHAPTER VIII.

NATIVE METALS... 219

CHAPTER IX.

ORES OR MINERALS YIELDING COMMON AND
USEFUL METALS 248

CHAPTER X.

USEFUL MINERALS BEING ORES OF THE LESS
IMPORTANT METALS ... 268

IN SEARCH OF MINERALS.

CHAPTER I.

MINERALS AND THEIR HOMES.

It is well in starting on a ramble in search of any object that we should have a clear notion of what we are looking for, and when the object of interest has reference to a department of Natural History this is especially necessary. Our earth presents at the surface a large amount of animated matter; plants and animals of many kinds, each living in its place, after its own instincts, each making use of, and made use of by, the rest; all flourishing within certain limits and increasing at a certain rate, the increase and the freeness of growth being governed by laws regulating life, which we can study, and to some extent understand. Most of these plants and animals live on the surface of the earth, or in the water which forms a part of the earth. All are surrounded by and enclosed within the atmosphere, which is also a part of the earth. We may regard the earth itself as non-living or dead matter, in contradistinction to living matter. Land, water, and air afford to living beings a mechanical support and medium consisting of non-living matter on which that which lives is so immediately dependant that we cannot conceive the existence of the latter without the former. It is usual to employ the words "organic" and "inorganic," in reference to these two conditions of living and non-living matter. We assume all that has life to possess *organs* or means by which life is carried on, and by which food is obtained and utilized, and whatever these means may be the structure so provided is "organized." That which is not so organized, though consisting of precisely similar atoms of the different elementary substances, has no life, and is inorganic.

We thus arrive at a knowledge of the great and essential difference between the plants and animals and the earth on which they live. It is the presence of life. Life is a mystery that no one has yet penetrated, a power recognised and vast, but utterly unknown except in its effects. It acts through matter, but greatly alters the condition of matter, evolving new combinations, and affected by new laws. Its presence or absence determines the conditions of the aggregation of atoms, and though we cannot draw a precise line between organic and inorganic nature in every case that is presented, we may generally feel and know where life has existed, although we cannot explain the exact conditions of structure.

Minerals belong to the inorganic world. They are formed frequently, and they increase rapidly, but they cannot be said to be born or to grow, in the sense in which we make use of these terms in reference to organic beings. The difference between organic growth and mineral aggregation is not always easily determined, but we may recognise it by considering that the simplest forms of organic existence increase by the addition of cells already living, whereas the mineral can only increase by the addition of inorganic atoms, whether simple or compound. This does not teach us what life is, nor even where life begins; but it shows why the animal or the plant and the mineral, though consisting of the same elementary substances in the same proportion, must develop differently from the very commencement. The plant or animal consists of mineral matter, and grows or increases in dimensions in a certain mode which indicates the presence of life. Vitality may cause mineral matter to become an essential part of an organized being, but no relation exists on the other side, for the mineral cannot thus act with regard to the plant and the animal.

It is evident that however largely and fully the earth, the water, and the air may be peopled, the mass of inorganic matter on and amongst which organized beings live must be enormously greater than the organic structures. A very large part of inorganic matter enters in no way into organic matter as a constituent, and it is difficult, perhaps impossible, to say in what way certain elementary substances are essential to life, or why other inorganic matter would not answer the same purpose. We must accept these and many other incomprehensible things, and may not hope to do more. The collection of ultimate atoms into elements is assumed to be true for the purpose of assisting inquiry. The proportions of each elementary group of molecules, if there is really only one kind of atom, are, at any rate, fixed, and the result is before us, but we cannot explain it. At present, chemists do not know enough about ultimate atoms to feel sure of the nature of the supposed elementary substances.

All inorganic matter is mineral matter, but it is convenient to distinguish some elements and some definite combinations of elements under the special name of *minerals*. Other and much larger collections of mineral matter having a less definite character are conveniently treated of by another term. "Rock" is the term that has been employed, and all rocks or important combinations of mineral matter consist partly of definite minerals, but largely of indefinite admixtures of mineral matter. Rocks form the great mass of the earth, and are to be found everywhere; and minerals, the more definite substances of which they are composed, are objects of great and general interest, which it will be the object in these pages to bring before the reader.

To render the matter more clear, let us take a case near at hand, and endeavour to work out the distinction. In any district of England there is almost sure to be found a *soil* in which the plants grow. There is below this a *sub-soil*, from which the soil has been derived; a *rock* beneath the sub-soil, which has its own local but well-defined character; and within this rock, or dispersed in the soil above, are distinct *minerals* capable of accurate description.

Let us suppose the district to be near London. Here the soil is derived either from surface gravels, or from the clay called London clay. Both gravels and clay are often mixed with sand, and the greater or less mixture of this with the clay below converts it into a sandy loam or leaves it a stiff clay. Far below the London clay is chalk, but where the clay is thick the influence of the chalk cannot be traced at the surface. In such a district the only minerals to be found would be those that have been carried from a distance and now form part of the gravel, and those which may occur in cracks and fissures of the clay. But in clay there are few minerals. The neighbourhood of London is therefore not a favourable locality for stony minerals, while metals and precious stones are still less likely to be found. But let us take an excursion by rail, and cross England from London in the direction of Cornwall. In doing so we cross rocks of the most varied character, some of them presenting not only many but very many kinds of minerals, and some consisting almost entirely of one mineral. The contrast is striking, and we shall see in future pages in what way it is instructive. A journey to Wales and across it would also give us rocks containing numerous rare and interesting minerals, though by no means such as we should find in the first trip. On the other hand, a journey northwards, from London through eastern England by the Great Northern railway to Peterborough, and thence through

Lincolnshire, eastern Yorkshire and Durham to the Northumbrian coast, although quite as great in point of distance and full of interest to the botanist and the entomologist, would be singularly free of interest to the collector of minerals. Along this line of country there is none of that altered or crystalline rock in which minerals of interest are chiefly found.

The formation of minerals is not a question of geological time, nor are they limited to old rocks. In England many, and indeed most, of the older rocks are more rich in this department of natural history than those of newer date. But it is not so everywhere. The mountains of Switzerland, and the changes consequent on their upheaval, are more modern than any of the older rocks of the British Islands, but their wealth in minerals is not inferior. The mountains of Central Asia are probably still more modern. The Andes and Rocky Mountains are the newest of all, and their upheaval is of the date of some of our youngest rocks, but they afford minerals in great variety and of equal importance.

While we draw a line between minerals and rocks, and recognise fully the difference between them, it must not be supposed that all minerals are definite chemical compounds, or strictly crystalline in their structure. Nature does not bring out results mechanically identical, even under circumstances that seem to us precisely similar. We must accept as similar minerals many that vary in external appearance, and note as essentially different many that are much alike. And thus the classification of these substances is difficult, and, owing to many reasons, unsatisfactory. The general mode of grouping it is proposed to adopt in these pages has no other foundation than convenience, and will not help the student who would master the science of mineralogy. To learn enough about minerals to take an intelligent interest in the subject, is a matter by no means difficult, and is generally desirable. To become an accomplished mineralogist falls to the lot of few, and needs all the aids of exact science, both mathematical and physical.

The great divisions we shall adopt are natural in a certain sense. How far that sense is the one most useful for the purpose of creating an interest in a little-understood pursuit it must be left to the reader to decide.

Minerals are noteworthy for many reasons : first, for their value intrinsically, as natural objects used for ornamentation ; secondly, for certain useful properties in reference to physical science ; thirdly, for industrial uses, whether as metals or earths ; fourthly, for their rarity, their curious mutual relations, and their peculiarities.

Besides these definite points interest attaches, and attention is called to them, for many other reasons which it would be tiresome here to allude to. It is enough to say that they are some of them social, and some relate to strange superstitions.

The value of minerals generally is now estimated in a very different manner from that adopted in former times, and the scientific estimation, as well as market price, of those especially valued, is based on other grounds. Colour was once thought important as a source of value, and was even made the basis of a classification proposed so lately as the last century by an ingenious French chemist. No doubt lustre and colour are among the very striking properties by which precious stones are recognised and compared. But both are qualities that can only be regarded as casual and variable. Other optical properties, though characteristic, are not adapted for general use. To the ordinary observer, hardness, specific gravity, crystalline form, and above all, chemical identity, are more direct, certain, and easily-recognised characteristics. Wherever the nature of the crystalline structure is traceable, that alone may be accepted as the best guide. Where it is absent, the reference to class must be made in other ways, and it is thus possible for some specimens which exhibit crystalline form, and others of the same material without it, to have a different place in any scheme of arrangement. We repeat that the method followed in these pages is one adopted in technical works, and must only be regarded as a convenient mode of bringing together those minerals whose manifest relation is such as to be at once recognised.

Many minerals are remarkable for their beauty of crystalline form, their singular lustre, their adaptation to ornamental purposes, their rarity, their colour and brilliancy, or other properties at once recognised by the eye. These are *gems*, or *precious stones*, and form several large groups full of interest.

Other minerals are obtained and used chiefly in

their condition as *metals*. Some are very valuable, as gold, and are found chiefly in a native state. These are called precious metals. Others, as silver, copper, and quicksilver, less valuable, are also found in a native state, but more frequently in combination as ores. The native metals form a group about which a great deal of curious and interesting matter is recorded, and of which the chief members are only found abundantly in a few localities.

Some of the most useful metals, as iron, lead, zinc, and tin, are never found in a native state as metals, but are very abundant mixed with earthy or other elements. These are *ores*. They include a great variety of minerals, and are widely dispersed and found almost throughout the world. Some of the metals found occasionally in a native state, are also abundant in combination and belong to this class.

Of earthy minerals some, not included among the gems but consisting of similar elements in various combinations, *silica* being the prevailing ingredient, are called *silicates*, and are extremely important. Those not regarded as precious stones and used for other than ornamental purposes form an independent group. Clay is one of the most familiar examples of rocks of this class. Volcanic, and what are called "metamorphic" rocks, are of the same kind, and contain a great variety of minerals belonging to this group. Other earthy minerals, of which lime and magnesia are the most familiar bases, are found to possess curious and important properties, and require some notice. They are for the most part pretty widely spread. *Carbonates* and *Sulphates* are the basis of the most abundant and important varieties of limestone minerals. Nitre and heavy-spar are examples of nitrates and sulphates of other alkaline earths. Salt is a chloride of sodium.

Under these principal headings we may bring together all the earthy minerals of general interest, with the exception of those consisting chiefly of the two elements, carbon and sulphur, both combustible. Of the former, the diamond is the crystalline representative. There are besides carbon minerals not precious stones, and these comprise graphite or black-lead, coal of all kinds, and the mineral oils or bitumens. Sulphur, when found native, and some of its combinations, also belong to this sixth group.

This general outline of the method to be adopted in describing minerals in the following pages seemed to be necessary to enable the reader to start fairly in the search for these objects of natural history. It is not intended to dwell here at greater length on the subject of classification, but so much it is desirable that the reader should understand.

Having now introduced to the reader minerals as definite objects of search, as distinguished from the rocks in which they are found, it will be worth while to consider briefly the general nature of their homes, especially as regards the more familiar examples of each of the groups into which we propose to bring them. By the homes of minerals we must understand sometimes real geographical localities, and sometimes the geological conditions either peculiar to or preferred by them. We speak of them as having preferences as we should of trees or plants, and not without reason. We do not, indeed, assume that they possess instincts or intelligence, but, like everything else in nature, they have a real and distinct place to which they belong, and, when found elsewhere, we may often trace the cause of this kind of migration, and ascertain the change of condition that has caused it. The more we look into the history and progress of what are sometimes called the laws of nature, the more clearly do we recognise a simple, well-ordered plan, the result of a high intelligence and a strict adaptation of every event that can arise to ensure the carrying out of a preconceived system.

Characteristics of Minerals.

Before commencing the account of the various minerals and their history, it will be well to explain shortly how it is they are recognised, and what are their various natures and properties. This is not, perhaps, the most interesting part of natural history, and is unnecessary in the case of plants and animals. All are familiar with those differences that enable us to recognise the subdivisions of the animal and vegetable kingdom, but in the case of minerals the principles adopted in grouping are less understood.

Minerals are recognised partly by certain peculiarities that can be determined by the senses, and partly by their composition. The former are capable of being understood by description; the latter require the application of chemical methods. Each is important, and about each a few words are desirable.

External form.—Many minerals occur usually in the crystalline form. It is therefore very important to have a clear notion of what is meant by this expression, and devote some attention to the nature and origin of crystalline structure, and the effect of transparent crystals on light, whether the light be reflected from or transmitted through them.

To do this it will be best to refer to some simple instance of rocks in which crystalline structure prevails, and thus endeavour to connect rocks with the minerals we have chiefly to deal with. A granite quarry in Cornwall or Aberdeen, or indeed in any part of the world where such a rock as granite exists, will afford abundant illustration, or, if the quarry is not at hand, a fragment of granite will be sufficient. The formation of ice on our window panes in winter, or the rapid evaporation of a solution of some salt on the surface of glass in summer, will answer the same purpose.

If we take the granite we shall find it made up of three minerals,—quartz, felspar, and mica. In any collection of minerals detached specimens of each of these minerals may be seen, and each will be found to present certain distinct forms by which it is characterized. In the quarry there will be found besides these many other minerals which we shall have to discuss. Among them might be rock-crystal, beryl, topaz, garnet, tourmaline, or Iceland spar. Whatever the mineral, it will be found to be referable to some definite form, and that it splits more or less readily in plane surfaces in certain directions, not splitting at all in others. The split surfaces will be smooth and shining, and by continual splitting, always parallel to the same surfaces, the crystal may be reduced at last to a mathematical form, which is that belonging to the crystal of the mineral we are examining, and no other. The ultimate crystal, whether obtained by this systematic reduction or presented at first, must be regarded as the nucleus.

It is true that in many cases this result is not to be attained without difficulty and some management, the right plane to split not being readily found, or the splitting not being easy, even when it is found. But the shape may generally be learned by the angle measured between two faces, which will give to the mathematician a clue.

All combinations of the same kind in nature, or in other words, all crystals of the same substance, present similar forms. Atom by atom the whole has been built up in some very definite way, and generally in the same way for the same substance. The building-up is the result of what are called polar forces, acting on the molecules or atoms, and inducing them to arrange themselves in accordance with certain laws.

Everywhere in nature we may observe this curious tendency of matter to arrange itself in definite forms when the atoms are allowed to act freely and take their own course. Sometimes this is the case when the body is kept fluid by heat for a long time, and only allowed to cool very gradually. Sometimes it is observable when the substances have been kept dissolved in water which has been removed very gradually by evaporation. If a little nitre be dissolved in water, and the water be slowly evaporated, the nitre molecules will arrange themselves in a certain form, and a minute crystal of nitre be produced. If, however, the evaporation be rapid, the process becomes confused and no defined crystal is formed. In time, by continued exposure, a large crystal may be obtained from a small one, the crystal growing as regularly and as surely as a plant or animal, and with more uniformity. The condition required for perfect crystallization is deliberation, and proper time being allowed there is no difficulty in obtaining crystals perfectly defined and transparent. Without ample time, however, the crystals are imperfect and confused in shape, more or less opaque, and often coloured.

The form of crystals is nowhere better or more beautifully illustrated than in the freezing of water. By lowering the temperature of water below a certain point, the tendency of the atoms to cohere in crystalline form is induced, and when this happens with the moisture of the atmosphere, the result is the production of snow. Among the ordinary flakes of snow that fall in winter it is not difficult to find some perfectly formed. Several of the small perfect crystals joining together, and always selecting points by which to attach themselves, the beautiful forms of twin crystal and stars shown in the annexed diagram are produced.

It is easy to obtain a very beautiful illustration of crystallization with common alum, dissolved in water and permitted to crystallize as the water evaporates. A string suspended in the water will serve as a starting point for the crystals, and they will be seen to grow rapidly as the water departs. In nature the same thing occurs with quartz or limestone. The crystallization once started, crystals attach themselves at

SNOW CRYSTALS.

every point. Distortion soon obliterates the typical form, but the angles between the similar planes are always the same, and some good crystals generally remain.

Owing to the way in which crystals are built up, and the fact that they consist of compound atoms arranged symmetrically, with intervals between them, the passage of heat, light, and electricity through minerals varies according to the mode in which their atoms are grouped. Thus, rays of light entering and passing through some crystals are altered in position, broken up into colours, or even transmitted in two different directions. These phenomena are spoken of as double refraction and polarization, and are the result of a quality of two-sidedness conferred upon the light thus altered. Many curious optical peculiarities are the result of this condition of crystallization, and the nature of the crystal may not unfrequently be discovered by its effects on light. The action of crystals on the passage of heat and electricity varies in like manner according to the mineral.

There are certain comparatively simple and familiar mathematical forms into which most minerals can be reduced by taking advantage of the property of cleavage, by which is meant the splitting in planes parallel to those on which the original or fundamental crystal was formed. The cube, the pyramid, and the rhomb are the principal solids from which in their various modifications most minerals are built up.

To make this clear in a popular way let us refer to the well-known form of the pyramid. The simplest pyramid is that whose base is a square and the sides all alike and all equal triangles. Besides this there is the pyramid whose base is a six-sided figure or hexagon, but the sides alike. A pyramid, however, may have the apex or top either over the middle of the base, as in the pyramids of Egypt, or not over the middle; and in this latter case the sides are unequal triangles. Two similar pyramids placed base to base form a third figure. To one of these figures almost all minerals are referred, but the complications of form are numerous and varied.

The first diagram on the next page represents calcite, a crystal of carbonate of lime, exceedingly common and very easily obtained. It is called Iceland spar when transparent and is useful as illustrating the curious optical property of double refraction. The figure is called a rhomb or rhombohedron, and gives its name to one of the principal systems of crystallography. By slight modifications several solids can be produced by cutting off the different angles or faces according to a regular method. Among these forms is the double pyramid and prism of

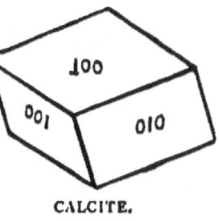

CALCITE.

the next figure, which is very often seen in quartz, and called Bristol diamond. These derivations of form are very important, as they afford a clue to many apparently irregular and almost shapeless forms of minerals, which are really only fragments of perfectly familiar crystals. The pyramid is thus seen to be only a derived form, and in its origin to be identified with the rhomb.

QUARTZ.

The diagram annexed represents a cubical crystal of salt. There are other cubical crystals, such as galena, fluor spar, and iron pyrites, of which illustrations will be found in another page. Although the cubical form is common to all these, there are important differences very clearly perceivable by examining the specimens. All these crystals are striated, that is, marked with lines parallel to each other. The crystal of salt is marked on each face with lines parallel to each side, so that there is a succession of

CRYSTAL OF SALT.

squares, each smaller than the outer one. The crystal of iron pyrites has lines all parallel to one edge, and at right angles to the lines on all the four sides adjoining. The crystal of fluor-spar has the striæ parallel all round the four sides, and these are at right angles to those at top and bottom.

Minerals sometimes take false forms; that is, one mineral is found assuming the shape of another. This is generally owing to the removal of a mineral once formed from the place in which it was formed, by the action of some natural cause, and its replacement atom by atom by another. It is only in a few cases that such changes take place. The replacing mineral is called a pseudomorph.

Colour.—Besides form, and of much less importance than it, minerals are characterised by colour. Colours are sometimes due to the actual composition of the mineral, and are part of its nature, but very often they are dependent on the presence of some foreign substance and are accidental. Thus, the colour of malachite is due to the copper which is a constituent part of the mineral, but the colour of emerald is owing to the presence of a minute portion of the oxides of chrome and some other metals. Play of colours, change of colour, iridescence or prismatic colours in the interior of a crystal, and opalescence, are all characteristic of certain minerals, and are valuable for identification.

Transparency.—This, in various degrees, is also sometimes characteristic, and may help to distinguish one mineral from another. Rock-crystal and Iceland spar are perfectly transparent. Alabaster is only *translucent*, permitting light to pass, but not clear enough to enable the eye to distinguish objects. *Opaqueness* is a quality of minerals through which no light or very little light can pass.

Lustre.—Of this property there are several varieties, all characteristic. *Metallic* is the lustre of metals; *vitreous*, that of glass. The terms *resinous*, *pearly*, and *silky*, sufficiently explain themselves. *Adamantine* is the lustre of the diamond. Intensity of lustre is called *splendent, shining, glistering,* or *glimmering,* according to its different degrees, and in each case there are minerals to which the term is specially adapted.

Optical Properties.—In transparent minerals a beam of light passing through is bent aside. This is called *refraction*, and the amount of displacement is eminently characteristic of the mineral. Sometimes the beam is divided and emerges as two beams, or an object seen through it looks like two. This is *double refraction*, and is also very characteristic. *Polarization* is a peculiar effect produced on light when passed through certain crystals. *Dichroism* is the term applied to the condition of a crystal in which different colours are seen when light is passed through it in two different directions.

Phosphorescence.—This property is possessed by some crystals which emit light either after being exposed to intense sun-light, heat, or electricity.

Fluorescence.—A peculiar property possessed by some minerals of transmitting one colour and reflecting another from a thin bed immediately below the surface.

Streak.—The colour shown by scratching a mineral with another that is harder. The mark made on the softer surface or the colour indicated by the powder scratched away are in some cases different from the colour of the mineral.

Stain.—The power possessed by some minerals of leaving a stain on paper or linen when drawn over them. It is very characteristic of lead.

Frangibility or Tenacity.—The following are regarded as the degrees of tenacity in minerals :—1. *Brittle*, or capable of giving off grains when scraped with a knife or file. 2. *Sectile*, yielding slices when cut, but not capable of being hammered without breaking. 3. *Malleable*, capable of being flattened under the hammer. 4. *Flexible*, capable of being bent without breaking. 5. *Elastic*, recovering its shape after bending.

Fracture.—When a mineral breaks so as to show curved concavities the fracture is said to be *shelly* or *conchoidal;* when it is smooth after breaking, it is said to have an *even* fracture ; when it shows sharp or jagged alterations it is *hackly*.

Cleavage denotes that a mineral can be divided mechanically, in certain directions yielding smooth faces, but if broken elsewhere it is uneven and jagged.

Hardness.—This property is tested by scratching the mineral with selected specimens of certain known and familiar minerals, of which the following is a list. These are numbered, and the hardness recorded by a number corresponding to those here given. (1.) Talc. (2.) Gypsum. (3.) Calc spar. (4.) Fluor spar. (5.) Apatite. (6.) Felspar. (7.) Quartz. (8.) Topaz. (9.) Sapphire. (10.) Diamond.

Toughness.—Resistance to breakage. Some minerals comparatively soft when tested for hardness resist breakage; others that are hard are easily broken. Some are both hard and tough.

Specific Gravity.—The relative weight of a mineral as compared with water affords a very ready means of determining some minerals. It is determined by first weighing the mineral in air, and then in water. The difference between the two weights will be that of the water displaced. The weight in air divided by the loss in weight (or the difference of the weight in and out of water) will give the specific gravity. The specific gravity of water is estimated at 1.

Taste.—This property may be used in some cases to determine minerals. Some are *astringent*, as sulphate of iron; some *sweetish astringent*, as alum; some *saline*, as salt; some *alkaline*, as soda; some *cooling*, as nitre; some *bitter*, as Epsom salt.

Odour.—This also varies in some minerals. Arsenic smells like garlic when any mineral containing it is struck or rubbed. The minerals containing selenium smell like horse-radish. Mundic or iron pyrites and some other minerals give out a smell of sulphur when rubbed or heated. Many minerals are bituminous; some, as quartz, are fetid when rubbed hard together. Some minerals emit an argillaceous or clayey smell.

Adhesion to the Tongue.—Some minerals when

touched with the moist tip of the tongue adhere slightly. This is a characteristic property.

Coldness.—The relative coolness to the touch of certain minerals, as quartz, is remarkable, and sometimes may be useful in determination.

Electricity.—Several minerals exhibit electricity on being rubbed, and a knowledge of this is useful to distinguish them from others that have no such property.

Magnetism.—This property is possessed by a small number of minerals, and very easily distinguishes them. Not only iron and iron ores, but nickel, cobalt, and platinum, attract the needle, and are magnetic.

All these physical properties are extremely useful and valuable in assisting the determination of minerals, but they do not indicate the actual composition. This can only be known by the help of chemistry. An easy and useful analysis can often be effected by the chemist with very simple means. Acids and other tests may be applied when the mineral is in its usual state, and the result will indicate its nature. Or by the aid of the instrument called the "blow-pipe" a sufficient heat can be raised from an ordinary wax-candle or spirit-lamp to melt small fragments, either alone or by the help of some substance which unites with the mineral and forms a kind of glass. The substances used for this are called *fluxes.* Borax is one of the most useful. Carbonate of soda and saltpetre are also extensively used. A piece of charcoal is required as a support; and with these fluxes, the charcoal, and a few simple implements, a great deal can be done to decide the nature of a specimen.

Some little knowledge of chemistry is, of course, needed to apply these means, but it may be easily learnt and practised by any one who takes an interest in the subject. A sharp eye, accustomed to observe slight differences, will enable any one to recognise at once a great many of the more abundant and important minerals; and the faculty of observation, cultivated to notice differences of form, colour, and other peculiarities, will be useful not only in this but in other departments of natural history. Minerals, however, even when of the same general nature and identical in mineral composition, put on very different appearances under different conditions, and must often be considered in reference to the part of the world and the nature of the rock from which they are obtained. Thus, to be familiar with them and recognise them, a study of their homes is desirable, and some facts about these we may now proceed to point out.

Homes of Minerals.

We have assumed that minerals, like plants and animals, have their homes, although we know that these homes are not always limited to one locality. Just as some plants prefer a particular soil and climate, but can with more or less ease adapt themselves to other soils and different climates, so minerals, though naturally belonging to, and developed in one rock, may often be found in others, and sometimes occur even where they are least expected. The real homes of minerals are, of course, the rocks in which they occur most freely, and with whose presence they seem to have some relation. Thus, a large number of beautiful and valuable minerals are found in granite; some, again, are almost confined to volcanic rocks, whether ancient or modern; some are only found among slates, others only where there is limestone. Some of these seem to have little to do with the material they inhabit, being found in cracks or cavities of the rock, and not connected with it by derivation. Thus, a common ore of lead (galena) is generally found in limestone. Other minerals consist of modifications of quartz altered from their original condition of sandstone, and now showing crystalline structure. These are almost always silicates. Minerals exist which prefer coal as a matrix, and a large number are found almost exclusively in altered argillaceous or clay rocks, such as slate.

As rocks of all kinds exist in various parts of the world, and are laid bare at the surface, or are reached by human labour, so minerals are widely spread and often have many habitats. Identity of mineral composition is not always certain to induce identity of form, although in this department of natural history there is perhaps little close resemblance of figure and function without identity of species. This is different from the case of plants and animals of distant countries and different climates, in which, as naturalists are aware, there is representation rather than repetition. The granite of Cornwall differs greatly from that of Aberdeen, and both of them differ from the granitic rocks of Charnwood Forest, in Leicestershire. Identical minerals are found, however, in each. The crystalline limestone of Derbyshire is different from the marble of Devonshire, but both contain the same kinds of ores of lead and zinc. The slates of Wales are quite unlike those of Cumberland, but many of the minerals are identical. And so, also, there are characteristic and even typical minerals in each case when similar rocks occur in widely-separated localities.

There is in minerals much of the same endless variety of what may be called personal character that is so remarkable in animals and plants; but in mineralogy, as in the other branches of natural history, this is only observable on intimate acquaintance. When we go into society and see a number of persons of the same race and the same class, with most of whom we are familiar, we notice this difference, and it is very rare even to find twin brothers that do not show some recognizable difference. But, if we happen to see together a few scores of negroes or Chinese, we think them so much alike that they can hardly be distinguished apart. With regard to animals, the difference seems still smaller. Who but a shepherd can know individually every one of a flock of sheep? And if we select animals of less high and complex organization, the difficulty becomes an impossibility with our limited organs of sight. Fancy knowing distinctively every ant in an ant-hill! But there can be little doubt that among ants there is as much personal identity as among Englishmen. It is only because our powers of vision and acuteness in detecting differences are limited and elsewhere directed, that we fail to see individuality throughout nature. There can be little doubt that even minerals bear some impress of their origin, and are sufficiently distinguishable if properly compared. With the microscope this may sometimes be done very effectually, and in some cases class characteristics are discovered due to the temperature and pressure under which the mineral was formed, and the presence or absence of water. Thus, notwithstanding general resemblances, there are special peculiarities of minerals, and the component parts of a rock, as developed in one place, are different from those of another locality. Besides this, there is often a real tangible difference among similar minerals, derived from the circumstances under which the rock was formed, changes which have been effected by the replacement of elements according to some subtle chemical process, or by the pressure of some disturbing force.

While certain minerals have, beyond all doubt, been formed by the evaporation of water from cold aqueous solutions of their elements or component parts, it is equally certain that many more are due to very slow changes taking place far underground, at high temperatures, and under great pressure. The natural formation of some minerals has been seen, and in others the method of nature has been so exactly imitated by art, that no difference can be detected in the result. With regard to many species, we have not yet attained to certainty as to their history, though in a general way we can indicate the transformations and substitutions that have taken place in their preparation. Others, especially those which contain many elementary substances, are more difficult to connect with geological conditions under which they might have occurred. Such minerals often occur in groups, and each group is associated with some peculiarity of rock structure.

India and the adjoining peninsula of Siam, and some of the large islands adjacent, are especially remarkable as being the homes of certain precious minerals, the whole class of Rubies being almost confined to that part of Asia. The Emerald and Beryl, though more widely spread than the Rubies, are chiefly common and abundant in South America. The Diamond has a still wider range, but has been only found abundantly in India, Brazil, and South Africa. The Turquoise is rarely met with out of Persia, though it appears there to be tolerably abundant.

Of the metals, the countries yielding gold are numerous, but large quantities have not yet been met with in the rock except under special geological conditions and in certain small districts. Among these the most interesting, although first discovered in modern times, are California and Australia. California yielded for a time enormous supplies. In Australia the discovery of gold gave the spur which led to the colonization of large districts. The annexed cut represents the locality where one of the earliest finds was made. The valley, whose river ran over golden sands, is seen studded with the tents of the diggers. The change thus induced in the country, and in Europe, was sudden and very remarkable. It

OPHIR, THE EARLIEST GOLD-FIELD IN AUSTRALIA.

is, however, only for a time that the accumulations of gold left behind in the gravels deposited by the rivers are available, and regular mining has not always followed those operations of digging and washing.

Tin is confined to a few spots; while Silver Lead, and Copper are found to some extent in a great variety of places. Other minerals, as Graphite, are met with only in small quantity, and in very few localities situated at great distances apart.

In most of these, and in many other cases that might be quoted, the homes of the minerals are either among mountains or among rocks which form the nucleus of mountain chains, or else are obtained from river sands and gravels of more or less modern date. There are, however, other minerals of great importance distributed more widely and in great abundance, and present under other conditions. Few, if any, countries are without some of the various combinations of iron with oxygen, carbon, and sulphur. None of any extent are without limestones and sandstones, and the minerals of which they are formed. The ores of metals may not be available or valuable, owing to the difficulty of treatment and the absence of cheap and plentiful fuel, but they are everywhere present. Ores of Copper, though less common than those of Iron, are often found in large quantity, and, although limited to certain rocks, are also present in most parts of the world. The island of Cyprus derives its name from the copper once found there in great abundance. Ores of Lead and Zinc are most common in the limestone districts, but also occur in metamorphic rock. They are rarely found without silver, and not often except in combination with sulphur. Mercury is only known in a few localities, and almost always with sulphur. Without bringing more instances forward, it is clear that some of the best-known minerals are not only limited in distribution, but exhibit a preference for certain parts of the world, certain associations with rocks, and also certain associations with other minerals.

There are, then, certain minerals, and those of great importance, whose habits are unsocial,

VIEW OF CYPRUS, THE EARLY HOME OF COPPER.

and which stand apart from their neighbours, and, having selected special homes, retain them and do not appear to migrate. There are others which are more social, some possessing this quality with reference to special earths, and many with reference to special metals; while some are altogether social, being widely dispersed over different countries and various rocks in combination with all varieties of rock and metal. The habits of minerals, as well as their habitat, are fair subjects for consideration, and the study will be found interesting and instructive.

Three great groups of earthy minerals may be characterised respectively as siliceous, calcareous, and aluminous. They are the most numerous, the most varied, and beyond all comparison the most abundant and common. There is no country, and, indeed, hardly any part of any country, in which rocks of each of these kinds may not be found. No doubt some arid tracts of sand, or patches of naked quartz rock, may be free both from calcareous and aluminous material. So, also, some wide tracts of limestone, equally naked and arid, are devoid of sand and clay. Perhaps, though much smaller even than these, some tracts may exist where aluminous minerals are present, forming rocks without silica, and without limestone, but they are quite exceptional. Almost everywhere, and indeed everywhere where there is animal or vegetable life, there is soil made up of these three ingredients, and occasionally minerals, as distinguished from rocks, indicating the presence of all three. Rock-crystal is one of a thousand forms of sandstone. Calc spar, or Iceland spar, is an occasional and unfamiliar result of the presence of limestone; corundum, whether as a useful mineral used in polishing (emery) or crystallized into the ruby or sapphire, is an exceptional condition of the alumina which forms the basis of all clays. Thus quartz minerals, limestone minerals, and aluminous minerals may be expected to occur, and are found, in all countries. Naturally the minerals of each class of rock might be expected to abound where the rocks are best characterised, but it is not always so. The rocks have generally been deposited in combination with water. The minerals found in them are due very frequently to changes that have taken place since the deposit, often consisting of the same elements as those that make up the rocks, but presenting them in new combinations, and collected into fissures or in open spaces produced by some contraction or upheaval of the rocks, or by some removal of a portion owing to the infiltration of water, or by a process of erosion the result of the passage of gases through the mass.

There will be much more to say with regard to the homes of minerals when we describe some of those that are most remarkable either as being very valuable or very rare. It is often by no means easy to detect a reason for the one condition or the other, either by reference to geographical or geological position; but we may safely assume that there must be some explanation consistent with what is known of the laws of chemistry and physics, and perhaps some day this, like so many apparently difficult problems, will be solved on more minute and careful investigation of the laws governing the arrangement of atoms and molecules in the formation of solids.

The familiar minerals that consist of single elements

cannot be many in number. In the first place there are but few elementary substances, and in the next place all, with a few exceptions, are rare, little known, and invariably found in combination with some more important metal or earth, and generally with oxygen also. All the simple elementary minerals are very limited in their distribution. Combinations of two elements are more numerous, but present no general characteristic, some being gems, some metallic minerals, and some salts. These are widely distributed. Combinations of three or four elements are still more numerous and abundant than these, and include a large proportion of those that possess general interest. But it is rarely the case that either these combinations, or simple minerals, are really exact. In nature it is much more common to find, in addition to the essential constituents, some substances accidentally present which greatly alter the appearance, while the proportion is too small to make an appearance in analysis. There are few minerals of much depth of tint in which metallic oxides are not the cause of colour, though the proportion is infinitesimal. Even the diamond, when coloured, and the ruby, owe their colour to this cause, while all coloured gems of high class and all the varieties of quartz are due to accidental admixtures with foreign material. It is, indeed, in very few cases that even transparent, colourless crystals are free from some quantity, however small, of other substance than the elements or chemical combinations of which they bear the name. The varieties of colour and accidental foreign ingredients are the causes of differences that characterise localities.

The mixture of chemistry and mathematics, involved in all accounts of minerals, cannot be ignored, and ought to be in some degree understood. No classification worthy of the name can be suggested that fails to introduce and even dwell upon these true causes of the peculiarities of different kinds, and without some system as a basis nothing can be stated beyond the baldest and least useful gossip. The system adopted need not be closely followed, nor need it be very minutely explained; but it should be made clear to such an extent that the reader will be able to gain some insight into the value of the various groups, and the extent to which they may safely be trusted. The peculiarities of minerals in all these cases indicate special homes, and are always highly suggestive.

CHAPTER II.

GEMS OR PRECIOUS STONES—THE DIAMOND AND RUBY.

ALMOST from time immemorial certain stones have been regarded as precious. This estimation was partly due, no doubt, to their rarity, but also to their natural beauty either as found in the earth or after being cut into crystalline forms and polished. Such stones possess certain properties of hardness, colour, or transparency, which add to their value. They have always been adopted for ornamental purposes, and have been valued at one time for superstitious purposes or for supposed medicinal properties. Some have been worn as amulets, some thought miraculous from their powers of attraction and repulsion; others as sensitive to the presence of animal or vegetable poisons, and others again as symbolic of mysteries in religion. Lastly, some were disliked and regarded as unlucky and even dangerous to those who wore them.

There are many ancient treatises on gems, most of them more remarkable for the superstitions they encouraged than for the clearness of their definitions. Indeed, there are few things more difficult than the identification of the names formerly given with the actual minerals they were intended to indicate. Pliny, the celebrated cyclopædist of antiquity, is one of the great classical authorities, and concerning ancient gems Theophrastus, another ancient author, has also written much and learnedly, if not very lucidly, on this subject. These ancient writers certainly applied the name of precious stones (λιθοι τιμοι) to minerals of very different kinds, and they still more frequently distinguished slight varieties of one stone by different names. We find confusion worse confounded when, as in the case of Pliny, an attempt has been made to classify. Few things can be less like than iron ores and the diamond, but both are by him called *adamas*, and this is only one of a multitude of similar instances.

Although convenient in a work like the present to consider in one group the various precious stones under the name of gems, and though such classification is not without the authority of great names, we must not claim it as of any real value. But no grouping of minerals is at present very satisfactory, and certainly none is very easily understood by the general reader. No apology, therefore, is necessary for the absence of technical grouping in this place. All that is meant here by gems is too well understood to need minute definition.

One curious application of the term is, however, worth mention. According to Pliny, the ancients included as gems all stones of beautiful colour, found in only small quantity and sufficiently hard to be useful when engraved as seals. With them the cutting of seals was in many respects the important use. With us gems are almost exclusively used as personal ornaments, such as rings, necklaces, earrings, brooches, pins, &c. In the East, however, even now rubies and emeralds are valued for their size and their intensity of colour rather than for their form, lustre, and freedom from flaw. In Western countries, under the aspect of modern civilization, perfect crystalline shape, purity of colour, brilliancy of lustre, and freedom from all internal blemishes, such as cracks, bubbles, or lines, are regarded as essential to a really valuable stone. Thus with us the cutting of precious stones is a process of the greatest nicety, and the most extreme precaution is taken to retain the natural faces of the crystal, while preserving the largest size and weight consistent with proper form and purity. In case of there being flaws or varieties of colour, great ingenuity is employed to conceal these in the cut stone and present to the eye only a perfect crystal.

Crystalline character is essential to most gems, and is the source of all that is most beautiful and most interesting in those that are transparent. Opaque gems, though sometimes of very beautiful colour and very valuable as curiosities, are rarely so effective as those which are transparent and crystalline. The most remarkable exceptions to this general rule owe their chief attraction to colour. Indeed, this may be regarded as their only claim to be regarded as gems.

Of the true gems the following are the most familiar and the most valuable:—Diamond, Sapphire, Ruby, Spinelle, Emerald, Beryl and Chrysoberyl, Topaz, Zircon (Jargoon and Hyacinth) and Garnets.

These form the first class, and with a few remarkable exceptions are the most valuable. Of the second class are the varieties of Rock Crystal, or vitreous quartz, including the Amethyst, the Cairngorm and the Aventurine; the varieties of Chalcedonic quartz, such as Cat's-eye, Mocha stone, Sard, Onyx, Sardonyx, Chrysoprase, Agate, and Carnelian; the varieties of Jaspery quartz, as Jasper and Heliotrope, or Bloodstone; and the varieties of resinous quartz, of which the Opal is so remarkable. Besides these varieties of silica, the Chrysolite, Turquoise, and some other stones are regarded as precious; and lastly, some substances of organic origin, such as Jet, Amber, Pearl, and Coral, take similar rank.

Jewellers employ some of the names above mentioned in a sense different from that recognized among mineralogists. Thus not only the true Ruby but the Spinelle, and even the Topaz, when of deep red colour, are all sometimes called Ruby; if green such stones are said to be Emeralds, if blue Sapphires, and if yellow they are called Topaz, without regard to their real nature and properties. There are also many dealers' names for several varieties of Garnet and other stones, which have been taken into general use, and which will be mentioned presently, but have no scientific accuracy.

Owing to the fact that the gems of almost all kinds are obtained from Asia, and were first known and used in that part of the world, and that the precious stones found in Europe, America, and Africa, are rarely so fine as the best of the Asiatic specimens, the term

"Oriental," or coming from the East, is often applied to the finest stones without much reference to their locality; thus Oriental Ruby, Oriental Topaz, Oriental Emerald, mean only that the stones are of the finest quality of their kind. So also the term *masculine* was applied by the ancients to highly-coloured stones, and *feminine* to those whose tints were more subdued. The terms are rarely so employed at present.

In the Middle Ages certain gems were regarded as symbolical. Thus, each of the twelve Apostles was represented symbolically, according to the following list:—

St. Andrew	*Sapphire.*
„ Bartholomew	*Carnelian.*
„ James	*Chalcedony.*
„ James the Less	*Topaz.*
„ John	*Emerald.*
„ Matthew	*Chrysolite.*
„ Matthias	*Amethyst.*
„ Peter	*Jasper.*
„ Philip	*Sardonyx.*
„ Simeon	*Jacinth* or *Hyacinth.*
„ Thaddeus	*Chrysoprase.*
„ Thomas	*Beryl.*

The ephod of the High Priest of the ancient Hebrews was ornamented with twelve stones representing the twelve tribes of Israel. These stones are supposed to have been engraved. They have been treated of in various works, and have been the subject of a special work, but little is certain about their meaning, nor are the stones themselves clearly determined.

According to Josephus, they were the following, the arrangement of the stones and the identification, with modern names, being subject to some doubt:—

1st Row.—SARD (red), TOPAZ (yellowish green), EMERALD (bright green).
2nd „ CARBUNCLE (dark red), SAPPHIRE (dark blue), JASPER (dark green).
3rd „ JACINTH or HYACINTH (orange), AGATE (black and white), AMETHYST (purple).
4th „ CHRYSOLITE (bright yellow), ONYX (blue and black), BERYL (light green).

They were engraved, no doubt, in hieroglyphics, and in all probability exist at the present day, perhaps among the treasures of Constantinople.

In the celestial vision described by the Apostle St. John, in Revelation ch. xxi., the walls of the New Jerusalem are built of courses of precious stone. The following is the list of these stones and their colours, as far as they can be identified. The arrangement is peculiar and interesting. It is not that of the rainbow, neither is it that of the order of the gems in the ephod described above.

NAME.	STONE.	COLOUR.
JASPIS	Jasper	Dark opaque green.
SAPPHIRUS	Lapis lazuli	Opaque blue.
CHALCEDON	Chalcedony	Greenish blue.
SMARAGDUS	Emerald	Bright transparent green.
SARDONYX	Sardonyx	White and red.
SARDIUS	Sard	Bright red.

NAME.	STONE.	COLOUR.
CHRYSOLITE	*Topaz*	Bright yellow.
BERYL	*Beryl*	Bluish green.
TOPAZION	*Peridote*	Yellowish green.
CHRYSOPRASUS	*Chrysoprase*	Apple green.
HYACINTH	*Sapphire*	Sky blue.
AMETHYSTUS	*Amethyst*	Violet.

It will be noticed that several of the ancient names do not correspond with the modern names of familiar minerals.

Planetary rings formed of the gems assigned to the several planets, each set in its appropriate metal, were regarded as of very great virtue in the Middle Ages. The following names of some stones and their settings, in reference to their astrological associations, may be interesting:—

SUN . . . . *Diamond* or *Sapphire*, set in gold.
MOON . . . *Rock crystal*, in silver.
MERCURY . *Magnet*, in quicksilver (probably amalgam).
VENUS . . *Amethyst*, in copper.
MARS . . . *Emerald*, in iron.
JUPITER . . *Carnelian*, in tin.
SATURN . . *Turquoise*, in lead.

Gems have been regarded as of great importance and high virtue in some medical ways, but little is now thought of the effect of powdered crystals, however great they are, or however rare they are. *Electuarium e gemmis*, *Confectio de hyacinthis*, or such-like compositions, are not now to be found in the Pharmacopœia as articles of the *Materia Medica*, and are not likely to be recommended by modern physicians, or to enter into modern prescriptions.

Gems are usually, though not always, crystalline, whether actual crystals or not. Sometimes they are found as crystals occupying cavities or open spaces in solid rock, sometimes they occupy veins or clefts in rocks, or occur between two rocks of different kinds. They are often associated with other crystals of less value, and they are only occasionally and exceptionally in such a state as to be fit for use as ornaments without artificial polish. Some valuable gems are only found rolled and water-worn, generally in gravel or the beds of rivers with their angles and faces worn and rounded, and in the case of the most valuable, the Diamond and Ruby, there is absolutely no clue to the matrix or rock in which, or with which, they have been formed. Others, among which are the Emerald and Topaz, are frequently met with in the rocks in which they were crystallized. When detached, the specific gravity of valuable gems being different from that of ordinary pebbles, they are naturally and readily separated from them by the action of moving water, and are collected into holes or corners, or are accumulated in greater abundance in some localities than others, owing to local and accidental conditions. With regard to the most valuable gems, the property of hardness is much more remarkable than that of weight. Even the Diamond, however, is nearly half as heavy again as rock crystal, or what are called Cornish or Irish diamonds, of the same size, and the Ruby is still heavier.

GEMS OR PRECIOUS STONES. 43

Although Asia, and chiefly the peninsulas of India and Siam, were in ancient times the principal known depositaries of the more valuable gems, and, with Upper Egypt, comprised almost all the countries that supplied the civilized world, it must not be assumed that they are exclusively privileged in this respect. Ever since the discovery of America, that vast expanse of western land, stretching almost from pole to pole, and having a range of almost continuous mountains from north to south, has been remarkable for its mineral wealth, not only in the precious and useful metals, but also in gems. The Diamond was soon found in Brazil and the Emerald in New Grenada, and these, as well as many other kinds, were introduced into Europe, the exploration of the deposits lasting for a considerable time. The Emerald and Beryl were at first found so abundantly in Central America as to make it almost doubtful for a time whether the stones of that kind known in the East before the Middle Ages had not possibly wandered thither across the Pacific. Europe also has supplied most of the other gems, some of them in large quantity, though not of the most valuable kind. South Africa has of late been recognized as the home of Diamonds, as well as gold, and Australia may have stores of gem-wealth, yet unknown in quantity, though not unindicated by specimens. It is only the Sapphire group that is still, as always, confined to the East, and of stones of this kind the supply was never large, and is now very small indeed.

In considering the gems we may conveniently

regard them as forming several principal groups, characterized by their respective hardness. The hard gems are, with few exceptions, the most valuable in the market, and are therefore the most highly appreciated for ornamental purposes. They are also remarkable for certain physical properties. By hard gems are meant those that are harder than the mineral called quartz, as known in the ordinary rock crystal. This beautiful, clear, transparent stone, exquisitely crystallized in a variety of forms, or found in large rock masses, is so common and so familiar, and is the parent of so large a proportion of the material of which the earth's surface is built up, that it is very convenient to make use of it as a dividing-line. We shall also see that it is very appropriate in other respects. Thus the quartz minerals form a second group. A third group includes a number of valuable stones softer than quartz; and there remains a fourth, comprising minerals directly derived from the animal and vegetable kingdom.

In the present chapter we shall confine our attention to those gems remarkable, not only for their beauty of colour and costliness, but for their hardness, and we take them nearly in the order of their hardness, that being also for the most part the order of their relative value. There is, however, one exception to this, inasmuch as the exquisite beauty of the colour and clearness of the Emerald gives it a value immediately after the Sapphire, and above other stones of greater hardness.

The Diamond.

DIAMOND NECKLACE.

Foremost of all amongst the glittering race,
Far India is the *Diamond's* native place;
Produced and found within the crystal mines,
Its native source in its pure lustre shines :
Yet though it flashes with the Brilliant's rays,
A steely tint the crystal still displays.
Hardness invincible, which nought can tame,
Untouch'd by steel, unconquer'd by the flame;
But steep'd in blood of goats it yields at length,
Yet tries the anvil's and the smiter's strength.
With these keen splinters arm'd, the artist's skill
Subdues all gems and graves them at his will.
Largest at best as the small kernel shut
Within th' inclosure of the hazel-nut.
Its fitting setting, so have sages told,
Is the pale silver or the glowing gold ;
And let the jewel in the bracelet blaze,
Which round the left arm clasp'd attracts the gaze.[1]

---

[1] This and other quotations in verse of a similar kind referring to many of the gems are taken from a Latin poem attributed to a certain Abbot Marbodus or Marbœuf, who lived in the middle of the eleventh century, and was in due time Bishop of Rennes.

The Diamond is by universal consent the queen of gems, the most brilliant, the most beautiful, and the most precious. It is the hardest substance known, and very brittle. Its name, derived from a Greek word signifying "unconquerable," indicates both this property of extreme hardness and that of being unaffected by the most violent heat of an open fire. It is also unaffected by any acid or alkali. In the East, whence it was first obtained, and where it has always been greatly considered, it was at one time believed to destroy poisons, to cure insanity, to act as a preservative against lightning, and to possess many other occult virtues. It is neither volatile under ordinary exposure, nor fusible, except by the intense heat of the voltaic arc, when it is converted into coke or graphite. It is thus proved to be nothing more than a crystalline form of carbon—that material which forms so large a part of all vegetable and animal tissue. No doubt the identity of this brilliant phosphorescent crystal, valued as the chief ornament of regal crowns, with the black stones of which nearly three millions of tons are weekly extracted from the earth in our own country, and as regularly burnt in our fires and furnaces, is a strange and startling statement. But there is no doubt of the fact.

The substance is certainly derived from the writings of Pliny and Solinus; and the poem is known as the "Lapidarium of Marbodus." It is both curious and interesting, and the knowledge of the Greek language it involves was not common at the period. The verses here given are quoted from a translation published in a well-known treatise on ancient gems by the Rev. J. King.

Coal, black-lead or graphite, and diamond, are three forms of the same substance, and that substance is the essential solid in all plants and animals.

To visit the haunts of the Diamond we must travel far, and into difficult and little-visited districts. India was celebrated from the most remote antiquity as the home of the Diamond. It was not till long after the discovery of America, and indeed only about 150 years ago, that Brazil began to be known as a diamond-producing country; and little more than ten years have elapsed since South Africa was first added to the number of the homes of diamonds. Let us endeavour to trace the history of the Diamond in each. There are, indeed, a few other places in which diamonds have been met with, but the stones have been few in number, and small in size and value. Among these the Ural Mountains, some parts of the country east of the Alleghany Mountains, in the United States of America, a small district in Algeria, and another in Australia, are the most remarkable. It is interesting to have found diamonds in these places, and the more so since they appear in each in similar combinations; but the discovery has been unimportant in any other than a natural-history sense.

On the south-eastern side of the peninsula of India were situated the two ancient kingdoms of Golconda and Visapur. They extended from Cape Comorin to Bengal, and the country, once a part of them, at the foot of a chain of hills composed of volcanic rock, enjoys the reputation of having yielded the diamonds that have rendered India remarkable. The Diamond-mines were spread over a wide tract, part of which was near the ruined city and fortress of Golconda, which is not far from the present city of Hyderabad.

About 250 years ago, a well-known French traveller, named Tavernier, after visiting all parts of Europe before he was twenty-two years of age, travelled for forty years as a dealer in precious stones, visiting Turkey, Persia, and the Indies, and in search of their valuable produce reached the Diamond-mines then open in India. There is no better description than his of these mines, which were at that time in full work, and had been so for a long period. They have since ceased to yield valuable stones. Nothing is known of the earlier discovery of Indian diamonds, and the following is a brief extract from Tavernier's account of the state of things in his day :—

"At Raolconda in the Carnatic, about 200 miles from the old city of Golconda" (probably to the southwest), "all round the place where the finest diamonds are found the earth is sandy, and full of rocks resembling those at Fontainebleau. In these rocks are several veins, half an inch to an inch wide, in which the miners insert little iron rods bent at the ends, and draw out the sand or earth, which they put into vessels. Among this sand or earth the diamonds are found; but as the veins are often irregular and hard, it becomes necessary to break up the rocks to follow the track. In this and in separating the diamonds a great lever is used, by which the diamonds are often cracked and flawed."

Another mine is described about 160 miles east of Golconda, in a plain near mountains. Tavernier states that at the time he visited this mine (about a century after its first discovery), it employed nearly 60,000 persons, men, women, and children: the men to dig, and the women and children to carry away the earth. The men dug ten, twelve, or fourteen feet, till they reached water. The earth was carried to a place prepared, and there carefully washed after being pounded with wooden mallets.

At a third place, near a town called Soumelpore, were washings of river-sands yielding diamonds. In all cases the method of working was the same, and no mining operations, properly so-called, were carried on. Similar localities are described in the island of Borneo, but they were not accessible to our traveller. In none of the cases do the diamonds appear to have existed *in situ*, but to have been transported from a distance, and deposited with the sand and gravel among which they had been consolidated.

The following curious account of the mode of trafficking in diamonds at Raolconda, at the date alluded to (the middle of the seventeenth century) is also given by Tavernier, and cannot fail to interest the reader:—" It is a pleasing sight of a morning to see the young children of the merchants and others, from the age of ten to fifteen or sixteen, all assembled under a great tree which is in the market-place; each with a quantity of diamonds in a little bag hung on one side, and on the other a purse fastened to his sash, in which some have from five to six hundred pagodas in gold. There they sit waiting till some person comes to offer them diamonds for sale either from that or any other mine. When anything is brought it is put into the hands of the oldest, who is regarded as the chief of the band; he, after examining it, puts it into the hands of the next, and so on from one to another till it returns to his own, without any one speaking a word. He then inquires the price of the merchandise, in order to purchase it if possible, and if by chance he buys it too dear, it is at his own loss. When evening comes these children collect together all they have bought in the course of the day, and, after examining the different stones, separate them according to their water, weight, and clearness; then affixing on each a price according with what they can sell them for to strangers, they, by comparing it with the price given, see what advantage remains to themselves. Lastly, carrying them to some of those great merchants who have always large assortments on hand, the profit is divided amongst them, only he that is the head or chief receives one-quarter per cent. more than the rest. Although so young, they are nevertheless such good judges of the value of stones, that if one of them should happen to purchase anything on which he is willing to lose half per cent., there is always one ready to give him the money; and, in offering them a parcel of stones, consisting perhaps of a dozen, they seldom fail to select four or five with some flaw or speck of defect in the corners."

It is said that the best stones have generally been

found near mountains, and that when they occur in marshy or damp ground, the stone is discoloured according to the nature of the soil. The first discovery of new diamond-fields seems always to have been accidental. To determine the value of stones Tavernier states that the Indians always examine them at night, placing a lamp with a large wick near a hole in a wall a foot square, and holding the stone between their fingers. He also adds that an infallible way to judge of the water is to carry the stone under a thick tree, where, by the shade of the verdure, they easily discover whether it is blue.

Very large diamonds have been found in this part of India, but the stones are now not sought for, and the mines are neglected. Other localities in India are in Bundelcund, near Panne, and on the river Mahanuddy, near Ellore, generally in the neighbourhood of trap-rock. In all the Indian diamond soils the stones are so dispersed that they are rarely found directly, even in searching the richest spots, because they are enveloped in an earthy crust, which must be removed before the gem can be recognized. In doing this the diamonds are often broken.

In Borneo and Malacca diamonds have been found in the river-sands, and the precious stones are said to be indicated by the yellow colour of the stony soil and a sprinkling of small black flints on the surface.

The diamond mines of Brazil were first worked in 1728. They are in ground which has a wonderful resemblance to that of the Indian diamond-fields, consisting of a kind of gravel locally called *cascalho*.

which is also the matrix of gold. It is apparently a surface deposit, part of a sort of pudding-stone common in the neighbourhood. This is removed from the beds of some of the rivers by diverting for a time the course of the water. Iron ores and jasper are among the stones found with the diamonds in this hardened gravel, and it is thought that the actual rock whence the diamonds are brought is a curious flexible quartz rock called *itacolumite*. Most of the Brazilian mines have since ceased to be worked, and in 1874 the produce had become very small.

The African diamond country was first discovered in 1867, in the Colesberg district. The mines are chiefly situated in the valleys of the Vaal and Orange rivers. There are two kinds of workings—river-diggings and dry-diggings. The former are deposits of gravel and drift, rich in specimens of jasper, chalcedony, carnelian, garnets, &c. The diamonds are thinly spread through the gravel, and are separated by washing. More regular mining operations are carried on in the dry-diggings, which are generally in circular pools of detrital matter, surrounded by shales. Here the diamond-bearing areas are of the nature of pipes, containing intrusive rock broken and reduced to a breccia, or pudding-stone, the diamonds being associated with garnets, fragments of rock, and magnesian minerals. Over the upper part of the pipes there is generally a deposit of tufa, in which are sometimes crystals of diamond. It is doubtful whether in any case the actual matrix has been found.

Australia was at one time expected to yield many

diamonds from the gold districts, but it has not yet been found important in this respect. It is, however, an admitted fact that diamonds have almost always been found in association with gold.

The largest Indian diamond on record is described by Tavernier as having been found in 1550, weighing uncut 900 carats, or 2,844 grains (nearly $6\frac{1}{2}$ ounces). It belonged to the Mogul Emperor in Tavernier's time, and had been reduced by cutting to 272·46 carats, or 861 grains. Its history in modern times is not known. There is a remarkable diamond of 367 carats in the possession of the Rajah of Moultan, in Borneo. It is said to be shaped like an egg, and to be of pure water and immense value.

The diamond called *Koh-i-noor*, now belonging to the crown of England, has a remarkable history. It was seen by Tavernier among the jewels of Aurungzebe, the Mogul Emperor, in 1665, and then weighed $279\frac{9}{16}$ carats, its original weight being differently described as $787\frac{1}{2}$ and 793 carats. Judging from its form, it has been supposed to be part of a larger crystal. When brought to this country it measured $1\frac{3}{4}$ in. in its greatest diameter, about $\frac{5}{8}$ in. in thickness, and weighed $186\frac{1}{16}$ carats. Being irregular in form and badly cut, it was recut, and thus reduced to $103\frac{3}{4}$ carats, but greatly improved in brilliancy and general appearance. It is thought that two very large diamonds have been taken from it since its first discovery.

According to the Hindu legends this stone was the subject of an heroic poem, and was worn by a warrior slain during a great war that took place about 2100 B.C. Its history is then lost for two thousand years, and we only know that, about the beginning of the Christian era, it was in the possession of the Rajahs of Malwa, from whom it was taken by the Sultan of Delhi in 1306. It remained among the treasures of Delhi till 1739, when it came into the possession of Nadir Shah. From his successors it was taken by Runjeet Singh, and remained in the Lahore treasury till given up to the Government of India on the annexation of the Punjab, and presented to the Queen of England in 1849.

This remarkable stone is said to have been turned up by a peasant when ploughing in a field forty miles from Golconda, and was then, in its rough state, as large as a hen's egg. It is now a fine diamond of good form, and takes rank among the most valuable gems in the world.

Among the very large and remarkable diamonds is one in the Russian crown, purchased by the Empress Catharine for a sum of £90,000, and an annuity of £4,000 per annum. It weighs 193 carats, and is as large as a pigeon's egg. The great Austrian diamond weighs 139 carats, but is rather yellow. It is valued at £100,000, and is said to have been bought for a trifle as a yellow crystal at a curiosity-shop at Florence. The Regent or Pitt diamond, belonging to France, weighs 136 carats, and is remarkable for its form and perfect limpidity. It is valued at £160,000. The blue diamond belonging to Mr. Hope weighs 177 carats, and, though an exceedingly valuable stone, is much less so in proportion to its weight than a colourless stone.

The largest of the Brazilian diamonds weighed when first found 254½ carats, and on being cut yielded a splendid brilliant of 125 carats. It is called the Star of the South. Another good diamond of this locality is in the crown of Portugal and weighs 120 carats, but most of the stones from South America are yellow and of inferior quality. South Africa has yielded several large stones of good quality. The "Stewart" is a remarkable stone, weighing originally 288¾ carats. It was found in 1872. This and other South African stones are well formed and large, but most of them have a tinge of colour, and they are rarely of pure water.

Diamonds with delicate tints of colour, though very beautiful, are not considered equal in value to those that are perfectly limpid and transparent. The blue diamond already alluded to, and others of decided hues of pink and green, often possess special interest and value, but those that are slightly coloured are second-class stones. Some diamonds are nearly or quite black. They are valuable for other purposes, but cannot be used as gems.

The diamond is highly electric, attracting light substances when rubbed, and after long exposure to the sun's rays becomes phosphorescent in the dark. It is always crystalline and often in twin crystals, the faces of the crystal being slightly curved. It is the angle formed by this curved face that forms the cutting surface so useful in cutting glass. It can be imitated artificially with the same result in softer material. The ultimate crystalline form of the diamond is eight-sided, the sides being all equal, but the derived forms are numerous.

Diamonds are weighed in *carats*, the carat being $3\frac{1}{8}$ grains nearly ($151\frac{1}{2}$ carats are equal to one ounce troy). By far the greater number of diamonds found and brought to market are very small. Out of 1,000 average stones weighed, half were under half-a-carat, and four-fifths of the whole weighed less than one carat each; of course large stones were excluded. When above ten carats and of good quality, stones are of so high a value that they cannot be estimated in the ordinary way.

The cutting and polishing of diamonds is an art supposed to have been discovered only in the middle of the fifteenth century. It was first carried on at Bruges in Flanders, and soon passed into the hands of the Jews of Amsterdam. The work is done by an iron wheel with the use of diamond dust, the wheel being rotated with extreme rapidity. Most stones of any size are reduced to at least one-half their weight by cutting, and great experience and judgment are required to produce the best result that the form of the stone and the flaws will admit.

There are three recognised modes or fashions of cutting diamonds, resulting in the production of *Brilliants*, *Rose Diamonds*, and *Table Diamonds* respectively. It depends on the natural form of the stone and the presence of flaws as to which is the most desirable in any particular case. The annexed

GEMS OR PRECIOUS STONES.

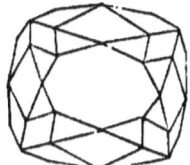

BRILLIANT.—TOP VIEW.

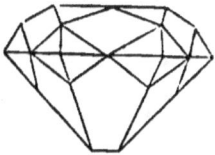

BRILLIANT.—SIDE VIEW.

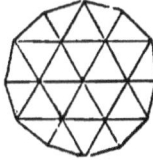

ROSE DIAMOND.
TOP VIEW.

ROSE DIAMOND.
SIDE VIEW.

diagram of varied cuttings will serve to render this more clear.

The Brilliant requires a thick rounded stone, nearly as deep as it is broad, and without flaws. It is cut with a large surface above, gradually smaller downwards, as seen in the cut, and is only available when the crystalline form is nearly perfect. This is the most rare and therefore the most valuable condition, and often involves much loss of weight to secure the greatest brilliancy.

The Rose Diamond may be described as half a crystal, the upper side being cut into a point above and flat below, as indicated in the diagram. It is less valuable than the brilliant, many stones not capable of forming good brilliants being capable of cutting in this form. Both Brilliants and Rose Diamonds are set *à jour*, so as to transmit as well as to reflect light.

Table Diamonds are flat and thin stones, and are often set with a foil which is intended to reflect light and thus increase the brilliancy. They are less valuable than other shapes.

The Diamond is extensively used not only as an ornament but for bearings and pivots in watches, for cutting glass, and of late for boring through hard rock, which is thus done with great rapidity. The difficulty of fixing the diamonds on the machine for boring is however very great, and many diamonds are lost in every operation. Inferior diamonds are used for these purposes, but the operation is still very costly.

" Diamond has the virtue of resisting all poisons, yet if taken internally it is itself a deadly poison." This notion was very prevalent during the Middle Ages, and is curious as being utterly unfounded in fact. The celebrated Benvenuto Cellini details an attempt to poison him in the Castel S. Angelo by causing diamond powder to be mixed with his salad, and attributes his escape to the rascality of the lapidary employed, who kept the diamond and administered pounded glass. It is also said that this was a poison administered to Sir T. Overbury in the Tower. Among other properties attributed to this gem it is said to baffle magic, dispel vain fears, and give success in law-suits. The same authority states that it is of service to lunatics and those possessed by devils, and repels the attack of phantoms and nightmare, rendering the wearer bold and virtuous.

Boort or *Bort* is a kind of Diamond of imperfect crystallization and generally of a spherical shape, but without regular cleavage. It is grayish white or blackish in colour, somewhat heavier than the ordinary Diamond, and not adapted for cutting. It is imported from the Brazils, forming sometimes ten per cent. of the rough diamonds from that country. Broken up and pounded in a mortar it is employed as a material for polishing other stones.

The Sapphire or Corundum Group.

Fit only for the hands of kings to wear,
With purest azure shines the *Sapphire* rare :
For worth and beauty chief of gems proclaim'd,
And by the vulgar oft *Syrtites* named.
Oft in the Syrtes midst their shifting sand
Cast by the boiling deep on Lybian strand :
The best the sort that Media's mines supply,
Opaque of colour which excludes the eye.
By Nature with superior honours graced,
A gem of gems above all others placed ;
Health to preserve, and treachery to disarm,
And guard the wearer from intended harm :
No envy bends him, and no terror shakes ;
The captive's chains its mighty virtue breaks ;
The gates fly open, fetters fall away,
And send their prisoner to the light of day.
Dissolved in milk it clears the cloud away
From the dimm'd eye, and pours the perfect day ;
Relieves the aching brow when rack'd with pain,
And bids the tongue its wonted vigour gain.
But he who dares to wear this gem divine
Like snow in perfect chastity must shine.

Under this name are included not only the Sapphire,

but the Ruby and some other gems, all of which when of good quality equal, and sometimes surpass, the Diamond in value, weight for weight, and are next to it in hardness. Their composition, however, is entirely different, being in fact not very different from that of the exceedingly common and familiar material, clay.[1] But these stones differ as much from clay in all recognizable properties as the Diamond does from coal. The gems of this group are chiefly natives of Asia, though specimens have been found elsewhere, but no variety of true Sapphire has been rendered more common by any large supplies from recently-discovered continents. The stones of this group are generally distinguished by jewellers as Oriental, not from their locality, though generally found in Asia, but marking their greater value, their hardness and their indestructibility.

The Sapphire is blue and the Ruby red, the colours in fine, well-polished specimens being the most exquisite in delicacy, richness, and depth that are presented in nature. The transparent, heavenly blue of the Sapphire, and the clear, deep blood-red of the Ruby, are recognized as typical of shades of colour that art may imitate, but can hardly rival.

The gems bearing these names are transparent crystals of corundum, the Kúrún of the Indians. Besides the two familiar colours there is a yellow-white or colourless variety called *Oriental Topaz*, and a green variety called *Oriental Emerald*. The true Emerald and the true Topaz are chemically very

---

[1] Corundum and the Sapphire are crystallized alumina. Clay is silicate of alumina.

different from corundum and have different optical properties. They also crystallize in different forms. Besides the varieties already mentioned, but somewhat more rare, there is a violet corundum called *Oriental Amethyst*, and another having a singular star-like reflection called *Asteria* or *Star-stone*. Some sapphires in which there is a pale red or bluish reflection are called *Girasol Sapphires*, and others with a pearly reflection, *Chatoyant* or *Opalescent Sapphires*.

Of Rubies, again, there are some varieties of colour and stones that are not corundum, which often bear the name of Ruby but with a qualifying title. The Spinelle Ruby and the Balas Ruby and their varieties are not corundum minerals.

The corundum gems generally terminate in six-sided prisms; they possess double refraction, and they become electric by friction. They are not acted on by acids and are rather improved than injured by fire, the red and yellow varieties being sometimes heated intentionally to render their colour more intense.

These stones, like other precious stones of the East, have been supposed to possess medicinal properties. Thus, they were said to purify the blood, to strengthen the system, to quench thirst, and to dispel melancholy. Used as a talisman, the Sapphire was believed to avert dangers and ensure honour and wealth.

The Sapphire has long been a symbol of investiture into the sacred office of bishop.[1] It is used in rings, and is generally surrounded by brilliants. The value of stones of great purity and good colour, and

[1] The Amethyst has been used to replace the Sapphire on some occasions for this purpose.

of moderate size, is greater than that of diamonds of the same weight. The Ruby was well known and greatly valued in ancient times, and is by some supposed to be the stone called *Sardius* or *Sard*, mentioned as the first among the twelve stones placed on the ephod of the High Priest of the Hebrews, and engraved with the name of Reuben. The engraving of the stone, if really a Ruby, would indicate a considerable knowledge of the arts, owing to the extreme hardness of the mineral, and the stone is more likely to have been the Sard of modern times.

The Island of Ceylon, and the kingdoms of Pegu and Burmah, contain almost the only Ruby and Sapphire-mines of the world, and the supply of stones has always been small, fine sapphires being especially rare. The Ruby-mines of Burmah are about sixty miles from the capital towards the north-east. The mode of working is described as merely sinking pits to a stratum called the gem-bed or ruby-earth, which lies at various depths, sometimes only two or three feet, and sometimes more than forty feet, from the surface. Often it is not found at all. It is an earthy sand, and when entered, drifts or lateral openings are driven in it, and the bed followed till it is exhausted, or till it is necessary to sink another pit. The thickness of the bed varies from a few inches to two or three feet.

The Rubies found are for the most part small, or if large are full of flaws. The Sapphires are found in the same earth, but are much more rare. When they occur they are however, of larger size and better quality than the Rubies. Well-formed crystals of Ruby occasionally occur, but the vast majority of the stones are rounded and water-worn, indicating a long period of river action.

There is reason to suppose that the blue Sapphire of modern times was the *Hyacinthus* of the ancients. It is thus described by a Latin writer, who lived about two centuries later than Pliny : " Among those things (cinnamon, &c.) of which we have spoken, is found also the Hyacinthus, of a shining sky-blue colour,[1] a precious stone, if it be found without blemish, but extremely subject to defects. For, generally, it is either diluted with violet or clouded with dark shades, or else melts away into a watery hue, with too much whiteness. The best colour of the stone is a steady one, neither dulled by too deep a dye nor too clear with excessive transparency, but which draws a sweetly-coloured tint from the double mixture of brightness and purple. This is the gem that feels the air and sympathises with the heavens, and does not shine equally if the sky be cloudy or bright. When put in the mouth it is colder than other stones. For engraving indeed it is by no means adapted, inasmuch as it defies all grinding ; it is not, however,

[1] Aeris ecce color tum cum sine nubibus aer.   OVID, *A. A.* iii. 174.

"The colour of the air we view on high
 When not a cloud is seen through all the sky."

The exact shade of the sky in the climate of Rome. — King's " Antique Gems," p. 49.

entirely invincible, for it is engraved upon and cut into shape by the diamond."

In spite of the difficulty of cutting, there are some exquisite antique gems cut in Oriental Sapphire. Among them is a Julius Cæsar, the stone an octagon, and of the finest deep colour. There is also a head of Phœbus, on a pale stone, nearly hemispherical, and a magnificent head of Jupiter, also on a pale stone, nearly an inch in diameter. Of the date of the Middle Ages is a portrait of Pope Pius III., on a beautiful Sapphire three-quarters of an inch square, regarded as an inestimable gem.

The Ruby, or red Sapphire, is believed to be the ancient *Lychnis*. It is thus alluded to by Pliny: "Of the same family of blazing stones is the Lychnis, so called from its lighting up by lamp-light, and of extraordinary beauty. The most esteemed locality is India. When warmed in the sun, or by friction with the fingers, it attracts straw and scraps of paper." He also speaks of it as shining with the red of the cochineal, and only surpassed in hardness by the Diamond.

Both the Ruby and Sapphire in ancient times, and even now in India, are often strung in an uncut state and used as necklaces. They are employed in large numbers in horse and elephant trappings. Some of these ornaments sent over to the Great Exhibition, in 1851, were covered with so large a number of rubies that it was almost impossible to count them correctly. Few of them however were unflawed, and none were polished.

Tavernier, in his account of his travels, states that rubies were occasionally found in Bohemia and Hungary. He describes these as enclosed in stones, which on being broken, sometimes, but rarely, contain gems, and states, that out of a hundred such stones presented by the celebrated Wallenstein to the Viceroy of Hungary, in the presence of Tavernier, only two contained rubies, but both these were of value.

Large Oriental rubies are scarcely ever without flaws, and the true ruby is not easy to distinguish from less valuable stones that imitate this gem in colour and external appearance. It is said that the King of Ava possesses a ruby as large as a small hen's egg, perfect in colour and water, and set as an ear-drop. Its value is inestimable, and far exceeds that of a diamond of similar dimensions. Several large but imperfect and flawed rubies have been described, some of which have been used for cameo or intaglio work, but fine stones of large size are excessively rare. Those of deep colour and good size have been called *Carbuncles*, to distinguish them from smaller rubies, but the name is now applied to varieties of garnet, much less valuable and of very inferior hardness, and would seem to belong properly to such stones. Only those rubies whose weight exceeds twenty carats were called carbuncles, and it is recorded that there were 108 of these in the throne of the Great Mogul. The largest known ruby is in the crown of Russia, and was brought from China, but its history is not known. Mr. King states that he knew only of three instances of antique sculptured rubies, and of these only two were on fine stones, both rather pale. Antique intaglios on rubies are even less common than on sapphires, and large rubies, like large sapphires, or indeed any transparent crystals of the Corundum group without flaws, are altogether exceptional. Some of the finest rubies have been spoilt by piercing, either before or after being cut. In some cases this has been done in India, but in others at a more recent period, and after having been brought to Europe.

Tavernier describes that in his time the finest rubies were obtained from Ceylon, and were saleable at almost any price, if exceeding five carats and perfect.

The variety of Sapphire called *Asteria*, or *Starstone*, is, when perfect, a wonderfully beautiful gem. Sometimes six, or even twelve rays appear, which change their place with the position of the stone. The condition seems to arise from some combination of conditions in cutting and structure, and can only rarely be obtained in perfection. The floating appearance of the star, only seen as if beneath the surface, adds much to the beauty of the effect and to the value of the stone, which is rare as well as beautiful.

The following rather extravagant lines are believed to refer to the Asteria, which was certainly called by the ancients *Ceraunite*, as there suggested. It is difficult to account for the superstition with regard to the origin of the stone, or its connection with thunder, but our forefathers rarely troubled themselves with accuracy in their poetical allusions with reference to such subjects. The two varieties of the stone are believed to be recognised:—

> When flash the levin-bolts from pole to pole,
> When tempests roar, when awful thunders roll,
> From clashing clouds the wondrous gem is thrown—
> Hence styled in Grecian tongue the *Thunderstone*.
> For in no other spot this treasure's found,
> Save where the thunderbolt has touch'd the ground:
> Hence named Cerunias by the Grecians all,
> For what we lightning they Ceraunus call.
> Who in all purity this stone shall wear,
> Him shall the bolt of heaven ne'er fail to spare;
> Its presence, too, protects from all such harm
> His city mansion and his blooming farm.
> Nor if he voyage o'er the briny deep
> Shall lightnings strike or whirlwinds whelm his ship.
> Thy foes in law, in battle, it confounds,
> And with sweet sleep thy grateful slumbers crowns.
> Two different species of this potent stone,
> Two different colours, are to mortals known:
> One, like the crystal bright, Germania sends,
> Which with its red and azure colour blends.
> The Lusitanian with the pyrope vies'
> In flamy radiance, and the fire defies.

The Green Ruby, or Green Sapphire (Oriental Emerald), is the *Smaragdus Scythicus* of Pliny, and is a rare and valuable gem, obtained by the ancients from Egypt, or more probably from Ethiopia. It is inferior in colour to the true emerald.

The aluminous gems have recently been obtained artificially by chemical processes, and very beautiful crystals of considerable size and great hardness are to be found in collections (see p. 90).

CHAPTER III.

PRECIOUS STONES, CONTINUED.— THE EMERALD, ZIRCON OR HYACINTH, TOPAZ, SPINELLE, TOURMALINE, AND GARNET GROUPS.

The Emerald Group.

Of all green things which bounteous earth supplies,
Nothing in greenness with the *Emerald* vies ;
Twelve kinds it gives, sent from the Scythian clime,
The Bactrian mountain and old Nilus' slime.
Highest *their* value which admit the sight,
And tinge with green the circumambient light :
Unchanged by sun or shade their lustre glows,
The blazing lamp no dimness on it throws.
But best the gem that shows an even sheen,
Lustrous with even never-varying green.
Of mighty use to seers who seek to pry
Into the future hid from mortal eye.
Wear it with reverence due, 'twill wealth bestow,
And words persuasive from thy lips shall flow,
As though the gift of eloquence inspired
The stone itself, or living spirit fired.
Hung round the neck it cures the ague's chill,
Or falling sickness, dire mysterious ill ;
Its hues so soft refresh the wearied eye,
And furious tempests banish from the sky.
If steeped in verdant oil, or bathed in wine,
Its deepened hues with perfect lustre shine.

THE true emerald and the beryl, with some other

gems less known to jewellers, belong to this group. The Emerald and Beryl are exceedingly different in their composition from the valuable gems described in the last chapter, but in value they rank next to the ruby and are certainly among the most beautiful of the precious stones.

The stones now under consideration consist of a combination of silica and alumina, with about 12 per cent. of an earth called *glucina*, the latter being very rare, and not known to exist except in crystals of this group. Specimens of the gem that have been analyzed show small quantities of chrome, magnesia and soda, and some other ingredients; and the colouring matter once thought to be derived from carbon, and to be identical with *chlorophyl*, the colouring matter of plants, is now known to be due to the oxides of these elements, chiefly chrome.

The crystals are usually prisms, often of some length, which are easily split at right angles to the axis. When first found emeralds are comparatively soft, and contain water, which is parted with on exposure, and the stone becomes harder. The hardness in the ordinary state is less than that of most of the valuable gems. The colour is lost on exposure of the crystal to low red heat, which renders the stone opaque and destroys its value.

Although much valued in India as an ornamental stone, it is doubtful whether any more than a few real emeralds have ever been obtained from that country. The principal modern source of supply is the eastern Cordillera of the Andes, near Bogotá, in the United

States of Colombia, at a place called Muzo, where the stones occur in a siliceous limestone containing fossils. Emeralds are also found in Peru. That they were formerly found in the East is probable, but owing to the doubtful identification of the true emerald with precious stones bearing the same name in ancient times, it cannot be regarded as certain. The word *smaragdus*, generally translated emerald, probably included several green transparent stones.

Like the other precious gems the emerald was once believed to possess many wonderful qualities. It was supposed to be good for the eyes, to serve (taken internally in a powdered state) as an antidote to poisons and the bites of serpents, and to cure dysentery, the plague, and infectious fevers. When worn externally as an amulet it was expected to cure epilepsy, to drive away evil spirits, and to prevent those provided with it from slavish fear of any event. In the East, it is still believed to avert bad dreams, to cure the palsy, and in many ways to give health and comfort to those possessing it.

The shape of the emerald is generally longer than broad. One of the most remarkable and largest stones known, however, is a crystal now measuring two inches in length and rather more than that across two of the diameters, the third being but little less. This stone belongs to the Duke of Devonshire, and weighs more than $8\frac{1}{2}$ ounces, independently of a small fragment of quartz to which it is attached. It has, however, many flaws. A smaller stone, but freer from flaws, is the property of Mr. Hope, and is valued at £500. Very much larger crystals are known, one weighing more than 12 pounds avoirdupois. These are from Siberia, but they are not fit for the jeweller. There is little doubt that many large stones have been split or sliced, several being made out of one stone, and many entirely spoilt, owing to the habit among the natives in the mines of New Granada of trying the stones by striking them with the hammer before accepting them.

Considering the extraordinary number of true emeralds brought from the East into Europe, many of them before the discovery of America, it is strange that some writers have asserted that they belonged only to that continent. The great source of supply in former times appears to have been Mount Zabara, in the vicinity of Coptos, in Upper Egypt, where the old workings were discovered and re-opened some years ago but without any profitable result. These stones, when true, were very highly esteemed among the ancients, although many other crystals, some of little value, and even green glass, bore the same name, and were esteemed. The finest stones were often sculptured, and a number of very beautiful works of art in this mineral are preserved in collections, and have been minutely described by Mr. King, in his work on antique gems. The following description of the emerald by Pliny is quoted by him:—" After the diamond and pearl the third place is given to the emerald for many reasons. No other colour is so pleasing to the sight, for grass and green foliage we view with pleasure, but emeralds with so much the greater delight, as nothing whatever compared with them equals them in the intensity of its green. Besides, they are the only gems that fill the eye with their view but yet do not fatigue it; nay, more, when the sight is wearied by any over-exertion, it is relieved by looking at an emerald. For gem-engravers no other means of resting the eye is so agreeable, so effectually by their mild green lustre do they refresh the wearied eye."[1]

Beryl is the name given to stones having the same composition as the emerald, and generally the same form, but of which the colour is not green. Some beryls are colourless, some are blue or yellow, some of two colours, and some iridescent. The finest are of a sea-green colour, and are sometimes called *aquamarine*. The crystals are often of considerable length, occasionally exceeding a foot, and the breadth across has been known to measure 5 inches. But these gigantic stones are of no value for jewelry.

> Cut with six facets shines the *Beryl* bright,
> Else a pale dulness clouds its native light;
> The most admired display a softened beam,
> Like tranquil seas or olive's oily gleam.
> This potent gem, found in far India's mines,
> With mutual love the wedded couple binds;
> The wearer shall to wealth and honours rise,
> And from all rivals bear the wished-for prize:
> Too tightly grasped, as if instinct with ire,
> It burns th' incautious hand with sudden fire.

---

[1] King's "Antique Gems," p. 35.

> Lave this in water, it a wash supplies
> For feeble sight and stops convulsive sighs.
> Its species nine, for so the learned divide,
> Avail the liver and the tortured side.

Beryls are far more common than emeralds. Gigantic specimens have been found in New England, in the United States, — one of them from Grafton, New Hampshire, weighing 2,900 lbs. This specimen is 4 feet 3 inches long, and 32 inches across in one direction. Another a foot long and nearly 4 feet across in one direction, weighs nearly half a ton. A stone weighing 80 lbs. is in the British Museum. These large stones have no value for ornamental purposes.

The distribution of beryls is very wide. Besides the United States they are found in the Brazils, and have been obtained from Hindostan and Australia. Specimens also occur in Sweden, in the Tyrol, in Bavaria, in the Isle of Elba, and near Limoges, in France. Small stones have been found in Cornwall, at St. Michael's Mount, and elsewhere, and others in several parts of Scotland. In Ireland very fine, well-coloured beryls, of good size and colour, and quite transparent, have been got from the granite of the Mourne mountains, and in Wicklow, besides other localities.

The Aquamarine, by which name the clear beryls were known to the ancients, is a beautiful stone, and is easily cut. It is often found of good size, and when the colour is sufficiently deep is valued for all kind of jewellers' work, and also for seals and intaglios.

It is a curious fact that *beryllus*, from which the name beryl is derived, is the low-Latin term for a magnifying glass, whence the German and French name for spectacles (*brille*). A writer in the middle ages, who died in 1454, also describes the beryl as "a shining, colourless, transparent stone, to which a concave as well as convex form is given by art, and looking through it one sees what was previously invisible." Mr. King remarks that probably the first idea of the invention of the magnifying-glass or lens was got by accidentally looking through a double convex clear beryl, or one cut *en cabochon*, a very usual form of ancient transparent stones. It was thence inferred that a clear piece of glass of the same shape might produce the same effect.[1] Before the discovery of America the beryl was a comparatively rare stone and was much used for sculpture.

Chrysoberyl, or *Golden Beryl*, is a beautiful yellowish-green stone, difficult to cut, but when of good quality and sufficiently large, almost rivalling the yellow varieties of diamond in lustre, polish, and colour. It is very hard and very heavy, much exceeding the emerald in both these respects. It contains nearly 20 per cent. of glucina. It is found in Ireland, in the Mourne mountains, and both in Brazil and Ceylon it has been found among the rolled pebbles in riverbeds. It has also been obtained from New York State near Saratoga, and in New England.

The chrysoberyl is only occasionally transparent,

---

[1] King's "Antique Gems," p. 40.

and varies in colour through several shades of pale green. It sometimes has a bluish opalescence; and when this peculiar opalescence seems to float, as it were, below the surface and in the interior of the stone, is called by another name, *Cymophane.* Stones of this kind are very well adapted for rings, and are much harder than Chrysolite, Moon-stone, or Cat's Eye.

Besides these are two more curious crystals in which glucina plays an important part. One, called *Phenacite,* is a silicate of glucina. It resembles quartz. Specimens of considerable size have been found associated with beryl in Siberia. They were either colourless or of a bright wine yellow, inclining to red. In nearly the same spot, and in rocks of mica slate, is found an emerald green variety of chrysoberyl called *Alexandrite.* When held between a light and the eye, it has a columbine red colour.

The *Euclase* (so called from the Greek εὖ, *easily*, κλάω, *I break*), though of comparatively small value to jewellers, owing to its rarity and brittleness, is a beautiful stone. It is hard, and takes a high polish, is of a pale, bluish-green colour, is transparent, possesses double refraction, and becomes electric by pressure. It contains nearly 24 per cent. of glucina, being much richer in this earth than the emerald. Originally brought from Peru, it has since been obtained from Brazil and from the Ural mountains, where it is associated with the emerald and corundum. In Brazil it occurs in chlorite slate, but in the Urals with gold-stream works.

Zircon Group, or Hyacinth.

There is a small group of gems remarkable for being composed in part of the earth called *zircon*, just as the gems of the emerald group are characterized by glucina. The root *zerk* is derived from the Arabic, and means precious stone. The crystals are very widely distributed, being found in our own islands, both in Scotland and Ireland; and they occur in Norway, Sweden, and Greenland; in France, Germany, and Italy; in Upper Egypt, and, in very fine specimens, in Canada. The crystals are of a peculiar brownish tint and adamantine lustre, and are doubly refractive. They are always found in old metamorphic rocks and volcanic districts. The zircons are among the minerals that have been made artificially.

The transparent bright-coloured zircons are called Hyacinth (or Jacinth) and Jargoon. The former are of various shades of red, passing into orange and poppy red; the latter either colourless or grey, with ill-defined tinges of green, blue, red, or yellow, generally cloudy.

> Three various kinds the skilled as *Hyacinths* name,
> Varying in colour and unlike in fame:
> One, like pomegranate, flowers a fiery blaze;
> And one, the yellow citron's hue displays.
> One charms with paley blue the gazer's eye,
> Like the mild tint that decks the northern sky:
> A strength'ning power the several kinds convey,
> And grief and vain suspicions drive away.
> Those skilled in jewels chief the *Granate* prize
> A rarer gem and flushed with ruby dyes.
> The blue sort feels heaven's changes as they play
> Bright on the sunny, dull when dark the day:

But best that gem which, not too deep a hue
O'erloads, nor yet degrades too light a blue;
But where the purple bloom unblemished shines,
And in due measure both the tints combines.
No gem so cold upon the tongue can lie,
With greater hardness none the file defy;
The diamond splinter to th' engraver's use
Alone its hardened stubbornness subdues.
The citron coloured, by their pallid dress,
Their baser nature openly confess;
With any kind borne on thy neck and hand
Secure from peril visit every land.
On all thy wand'rings honours shall attend,
And noxious airs shall ne'er thy health offend;
Whatever prince thy just petition hears,
Fear no repulse, he'll listen to thy prayers.
Midst other treasures to adorn the ring
This gem from Afric's burning sands they bring.

In spite of the authority of the poet, true and valuable hyacinths, though they may have been once found in Upper Egypt, are now chiefly met with in Ceylon, in the gravel of some streams, and are sold by the natives as inferior rubies. Indeed the name hyacinth is believed to be derived from the Persian and Arabic word *yacut*, which means ruby, and appears to have been applied by the ancients to some of the red sapphires. The term *jacinth*, identical with hyacinth, is frequently given to the cinnamon stone, a variety of garnet; but it only properly belongs to the stone now under consideration.

Though not worn much at the present day, this stone is valuable, and makes a superb ring-stone, if bright and free from flaws. It is said to deepen in colour on exposure to the air, reassuming its paler tint when kept long in the dark. It has rarely been found large and at the same time perfect. The above extract from the "Lapidarium" of Marbodus mentions the properties it was supposed to possess in the middle ages; but it is to be feared that the want of faith so characteristic of the present day will interfere sadly with the performance of much of this wonderful promise.

The Jargoon or Yellow Jacinth is probably a stone described by Theophrastus and other ancient writers as the *Lyncurium*, believed to have been formed out of the secretions of the lynx. The following quotation from the "Lapidarium" will point out its supposed properties, which appear to be quite as remarkable as, though quite different from, those of the hyacinth:—

> Voided by lynxes, to a precious stone
> Congealed, the liquid is *Lyncurium* grown;
> This knows the lynx, and strives with envious pride
> 'Neath scraped-up sand the precious drops to hide.
> Surpassing amber in its golden hue
> It straws attracts, if Theophrast says true:
> The tortured chest it cures, their native bloom
> Through its kind aid the jaundiced cheeks resume;
> And let the patient wear the gem, its force
> Will soon arrest the diarrhœa's course.

There are many beautiful ancient engraved stones of this gem, but all show a peculiarity which greatly detracts from their value, the surface presenting a worn and scratched appearance, worked in a curious manner, totally different from that of other stones, and due to its nature and certain difficulties in cutting. A peculiar porous texture is characteristic of the stone.

It is doubtful whether many sculptured gems described under this name were not really other stones. The tourmaline appears to be the mineral alluded to in some of the descriptions. The stone called *Lychnis* by Pliny, believed to emit light at night, is another; but there can be little doubt that not only in regard to this, but many minerals, accounts professing to be accurate descriptions of stones are really quite untrustworthy.

The stones of the Zircon class, except perhaps some from Canada, are now hardly regarded as precious in the jeweller's sense of the word. Besides the transparent crystals here described, there are other minerals, more or less opaque, containing the same peculiar earth, but not in any way available for ornamental purposes.

The Topaz Group.

From seas remote the yellow *Topaz* came,
Found in the island of the self-same name ;
Great is the value, for full rare the stone,
And but two kinds to eager merchants known.
One vies with purest gold, of orange bright ;
The other glimmers with a fainter light :
Its yielding nature to the file gives way,
Yet bids the bubbling caldron cease to play.
The land of gems, culled from its copious store,
Arabia sends this to the Latian shore ;
One only virtue Nature grants this stone,
Those to relieve who under hæmorrhoids groan.

The gems of this group are fluo-silicates of alumina. They are found in the British islands, in England, Scotland, and Ireland, generally in granite. The chief supply is from Brazil, but fine specimens have been obtained from Siberia, and exceedingly fine crystals from Tasmania. They are also found in India, Ceylon, and Pegu, in various parts of Australia, and in some parts of Germany. The name, as indicated above, is derived from an island in the Red Sea, from which the stone was obtained in classical times. It is, however, believed that the crystals formerly found in Topazion and called Topaz by the ancients, were chrysolites, and not what *are now called topaz.

The colour of the topaz is generally a beautiful wine yellow, and the stone is moderately hard. There are three well-marked groups :—1. The Yellow Topaz, of which the colour is remarkably modified by exposure to heat; 2. The Blue Topaz, or Brazilian Sapphire; and 3. The White Topaz, which when fine approaches the diamond in brilliancy and water. These varieties all include very beautiful stones, only to be distinguished from the diamond and corundum gems by their inferior hardness.

The yellow topaz should be of a rich, full colour, with a tint almost warm. Detached crystals come from Villa Rica, in Brazil, whence also we have pebbles capable of being cut into fine stones. In any state the colour of these Brazilian specimens can often be changed from yellow to pink or pale crimson by a very simple process, the stone being wrapped closely in amadou bound round it tightly with tin wire. The amadou is then lighted and burned, after which the colour is found to be changed, and only the surface polish injured. The experiment is hazardous, as the topaz is apt to crack and become flawed while burning; but the resulting stones are increased in value, and may pass as Balas rubies.

A variety, called the Saxon Topaz, is of a paler yellow, and when heated the colour is discharged instead of being deepened. These are much less valued than the Brazilian stones. Topazes from Mexico resemble the Saxon variety, and have little value. On the other hand, the Siberian crystals are generally good, and, having a bluish-green tint, resemble the aquamarine. Green and blue varieties of this kind of topaz are obtained from Kamtschatka.

The Blue Topaz is a very beautiful stone, and generally comes from Brazil. It is found of moderate size, up to 3 ounces, and when set with great care and full reference to its crystalline form is a valuable gem. It is sometimes called the Brazilian Sapphire, and it somewhat resembles the true sapphire in its full blue colour.

The White or Colourless Topazes, whether obtained from Tasmania, Brazil, or Siberia, are extremely beautiful, and approach the diamond in effect when well cut and mounted. The purest stones are sometimes called *Pingos d'agoa* (drops of water), from their extreme limpidity, and they are often cut as brilliants with an open setting.

The English topazes are found in Cornwall, at St.

Michael's Mount and elsewhere; the Scotch, near Cairngorm, whence very beautiful pale-blue specimens have been taken. The Irish are from the Mourne mountains.

Very large topazes have been found in Russia. There is a specimen in Paris weighing more than 4 ounces; and Tavernier mentions a stone in the possession of the Great Mogul, which weighed 157 carats, and is said to have cost £20,000 sterling. Generally, however, the value of the topaz is not very considerable.

Spinelle.

A group of gems comes next in order, comprising the Balas Ruby, the Rubicelle, and a number of other minerals, collected under the general name of *Spinel*, or *Spinelle*, and consisting of alumina and magnesia, generally coloured by a minute proportion of oxides of chrome or iron. Resembling in appearance the ruby, but of inferior hardness, different crystallization, and different specific gravity, the stones thus characterized are of comparatively small value, a fine stone of 24 to 30 carats not being worth more than from £10 to £15, whereas a fine ruby of one-fourth the weight is almost inestimable.

There are four varieties of colour in this group, the scarlet being called *Spinelle Ruby*, the rose-red, *Balas Ruby*, the orange-red, *Rubicelle*, and the violet, *Almandine Ruby*. There are not less than twelve distinct minerals, varieties of the spinelle, but not requiring notice here.

GEMS OR PRECIOUS STONES.

These stones are found chiefly in Ceylon, but they occur in other Eastern countries, and also in the northern parts of North America. The following account of the workings will be interesting. It is from a Persian author :—

"The mine of this gem was discovered after an earthquake in Badakshan had rent asunder a mountain in that country, which exhibited to the astonished spectators a number of sparkling pink gems of the size of eggs. The women of the neighbourhood at first supposed them to possess some dyeing property, but finding that they yielded no colouring matter they threw them away. Some jewellers suspecting their value gave them to lapidaries to be polished, but at first this was found to be difficult, owing to their want of hardness. It was afterwards effected with marcasite.[1] At first the stones were thought to be of great value, but when found so easy to cut were less esteemed."

The spinelle is the most valued of these stones, and after it the Balas ruby, but all vary in this respect according to the freedom of the stone from flaws and its size and crystalline form, and none approach in value the harder gems.

There is a variety called *Pleonaste*, opaque and of a dark or pearly-black colour, but having a brilliant lustre, which is sometimes cut and used for ornamental purposes. The other varieties do not rank as gems.

Iron pyrites.

Tourmaline.

A number of minerals are included under this name, all of them consisting chiefly of silica and alumina, with the addition of magnesia, the earth called lithia, or iron. Generally of an olive-green colour, often nearly black, tourmalines show every variety of transparency, from perfect clearness to opacity, and many varieties of brown, blue, green, and red tint. Brazil yields a bright-green variety. When transparent, the crystals exhibit the peculiar property of *dichroism*, appearing of a very different colour when looked at and when looked through. They are often red one way, and green or blue the other. They become curiously electric when heated.

It will readily be understood that the most transparent tourmalines are those most fitted for the use of the jeweller. On account of their deficiency of lustre, and their smoky or muddy tint, they are not held in great estimation as gems, but when well selected, cut thin, and set with a proper foil, they are sometimes very beautiful. They are more highly valued in Brazil than in Europe, and are there worn in rings, chiefly by dignitaries of the Church.

The *Rubellite*, or Red Tourmaline, when free from flaws, is a fine stone. There is a noble specimen in the British Museum of uncommon form and dimensions, presented to Colonel Symons by the King of Ava. It is valued at £1,000 sterling, and the colour is deep. The stones of this variety come from Siberia and Ceylon, as well as Ava. The others are chiefly from Ceylon and Brazil.

The electric properties of tourmaline are quite as remarkable as its optical properties, for they are not only more powerful than in other stones, but have relation to a curious condition of structure. The crystals of this mineral are, in fact, differently terminated instead of being similar throughout. The result is, that a prism heated in a particular way becomes positively electrical at one end and negatively at the other. When rubbed also it becomes positively electrical; and if heated it becomes electrical on cooling, the positive electricity being developed at the end where the number of facets is largest, and the negative electricity at the other. This condition is reversed by extreme cold. If a prism is broken while electrical from heat, the fragments present opposite poles, like artificial magnets. The crystals are often parti-coloured, being of one colour at one end and quite a different colour at the end opposite.

Tourmaline is common enough in many countries, but generally in the opaque condition. Black Tourmaline is called *Schorl*, and is abundant in granite, but of no value as a gem. Tourmalines contain boracic acid to the extent of from six to eight per cent. Their hardness is greater than quartz, but the difference is not considerable.

Garnets.

The Garnet group of gems comprises not less than six varieties of mineral composition, almost all silicates of alumina and some other base. The other bases are lime, magnesia, iron, or manganese. Besides these four, silicates of lime and iron, and silicates of lime and chrome, complete the list.

GROUP OF GARNETS.

The garnet varies in colour according to its composition, but red, brownish-red, and black are most common. The stones vary also in transparency, fracture, and size. When larger than a hazel-nut, they are hardly ever free from serious flaws. The name is derived from the Latin word signifying the scarlet blossom of the pomegranate-tree. The stones met with in commerce are generally from Bohemia, Ceylon, Pegu, or Brazil, but they occur in the hill districts of Cornwall and Cumberland, and are met with in Scotland and Ireland. Their hardness is inconsiderable compared with the other hard gems, and though the fine crystals are harder, the earthy varieties are less hard than quartz.

To be regarded as gems, garnets should be blood red, or of the colour of red wine, but these tints are often mixed with blue, so as to present shades of crimson, purple, and reddish-violet, or tinged with yellow, and passing into orange-red and hyacinth-brown. Jewellers class garnets as Syrian, Bohemian, or Cingalese, not so much with regard to their origin as to their relative fineness and market value. The most esteemed are those called *Syrian*, but properly *Syriam*, as they have nothing to do with Asia Minor, the name being derived from the capital of Pegu, whence the best stones have been obtained. There is a large supply of this kind of garnet brought from mines at Zöblitz in Silesia.

Garnets were largely employed by the Persians, and afterwards by the Romans. By the former the garnet was regarded with especial favour, as a Royal stone, during the later period of Persian history, and it frequently bears the sovereign's image and superscription. By the Romans it was sometimes used for sculpture, but not frequently, as very few fine works cut in stones of this kind are to be found in the different collections.

Cinnamon-stone is the name given to the best of the Lime-garnets. This stone is sometimes translucent, but rarely transparent, and it has a clear cinnamon-brown tint. It is found chiefly in Ceylon, and when transparent and of good colour is used as a gem, often replacing the hyacinth in the nomenclature of jewellers. Very good stones have been found in the United States, and in one mine in Maine, New England, there are cinnamon-stones of a yellow colour, and very beautiful. This variety is harder than quartz, but scratches it with difficulty.

Grossular is the name of a variety, so called from its

resemblance in colour to the green gooseberry. It is found in Siberia. *Topazolite* is an amber-yellow or honey-yellow garnet, found in Piedmont.

The Iron-garnets include some of the most valuable kinds, among them being the *Almandine*, also called the Precious Garnet, as well as the Common Garnet. The Almandine, when of good size, well-coloured, transparent, and without flaws, is a very valuable gem, and seems identical with the carbuncle of Pliny. Its colour is very deep, and it must be cut thin to be transparent. It is often found in sand and alluvium, but, like all the garnets, is derived from gneissic rocks. Ceylon, Pegu, Hindostan, Brazil, and Greenland are all countries where it has been found. It is also met with in Bohemia, where many garnets and other precious stones occur.

The best of the manganese garnets are called *Spessartine*. Their colour is deep hyacinth red, but they have little value.

There are two other varieties of garnet in which the alumina is absent. The first of these is the iron-lime garnet, which includes the Aplome, Melanite, and another. The *Aplome* is a deep-brown stone, the *Melanite* a velvet-black; both are opaque. The latter are found at Frascati, near Rome, and are known as Black Garnets. They have little value as gems.

The Lime-chrome Garnets — the last group — are called Pyrope, and include the *Carbuncle*. The name Carbuncle is applied to several stones, among them being the Almandine mentioned above, and a variety of ruby described on page 65. The *Pyrope*, or *Fire-Garnet*, is a dark-red kind, not often in good crystals, generally transparent, and frequently found detached from its matrix. Its colour is full crimson red, approaching that of a ripe mulberry, and resembling fire. For these reasons it was called by the ancients *anthrax* (ἄνθραξ) or coal.

> The *Carbuncle* eclipses by its blaze
> All shining gems, and casts its fiery rays
> Like to the burning coal; whence comes its name,
> Among the Greeks as *Anthrax* known to fame.
> Not e'en by darkness quenched its vigour tires,
> Still at the gazer's eye it casts its fires;
> A numerous race, within the Lybian ground
> Twelve kinds by mining Troglodytes are found.

The above account of the carbuncle, though imaginative, points to its most remarkable characteristics. When perfect and of good size it is a valuable as well as a very beautiful gem, and was formerly more highly esteemed than it is at present. The ancients obtained it from Carthage and Massilia. It is now found in Ceylon, in alluvial deposits, associated with more valuable stones. It is also obtained from Saxony and Bohemia.

The carbuncle possesses a certain amount of historic interest, a pendant carbuncle to the necklace of Mary, Queen of Scots, having been valued at 500 crowns, a very large price for those days.

As an ornament the garnet depends much on fashion. In the time of Charles II. a fine set of these stones was considered a magnificent ornament for ladies of the highest rank. During the last century they were less valued. They have since been more appreciated, but are still less considered than some stones of inferior beauty and hardness.

Although hard the garnet is not difficult to work, and it has been used for intaglios from time to time. The colour of the polished stone is less bright and clear than that of the ruby or even the spinel.

Coarse garnets are sometimes used as a substitute for emery in the work of preparing gems for polishing, and also for cutting, but they are of very inferior hardness. There are many rocks in which crystals of garnet are so extremely abundant as to be characteristic, but these are not often stones fit for the jeweller.

Before leaving the subject of gems it will be well to say something of imitative and artificial gems, which are much more common, and more frequently mistaken for really precious stones, than most people are aware. The manufacture is ingenious, and the result to the eye very wonderful.

All imitative gems are varieties of glass, and their basis is a paste or glass technically called *strass*, composed of silicate, potash, borax, the oxides of lead, and sometimes arsenic, all in a state of extreme purity. The proportions vary, each maker having a favourite and secret process; but in all cases a satisfactory result is only obtained by the most extreme care, not only in reference to the ingredients, but every implement and contrivance made use of. The crucibles employed must be capable of withstanding the strongest heat, and be impenetrable to the metallic oxides. The materials must be finely powdered and sifted, and thoroughly mixed; and the fuel should be wood, thoroughly dry, and cut into small pieces.

When a good strass is obtained the production of the imitative gem requires still further care and attention. To imitate diamond, white-lead in its purest form is added; for topaz, glass of antimony and purple of cassius; for sapphire, oxide of cobalt; for emerald, oxides of copper and chrome, or cobalt and chrome; for amethyst, oxide of manganese, cobalt, and purple of Cassius; for aquamarine, oxide of cobalt and glass of antimony; for precious garnet, antimony, manganese, and purple of Cassius; for rubies, a more complicated mixture is required, and powder of some imitative gems added to the ingredients.

True gems have been produced by chemical agency from the actual elements of which their crystals are formed in nature. Very remarkable specimens of sapphire have been obtained in this way. The specimens hitherto have been chiefly of the Corundum class, but minute crystals, believed to be diamonds, have lately been manufactured.

It is remarkable that both in the imitation and manufacture of gems the French have always taken the lead. So perfect is the imitation that a high reward was offered to any one who could pick out the false from the real of a group of stones sent by Paris jewellers to the London Exhibition of 1865. The challenge was not accepted.

CHAPTER IV.

Quartz Gems.

The position of Quartz among the gems has already been alluded to, and is rather peculiar. Silica, the name of the earth of which quartz is the crystalline form, is one of the most abundant materials of which the outer crust of our globe is composed, and it may be looked for in the sands of every seashore, and in every quarry where sandstone is worked. It is indeed present in almost every rock that does not consist exclusively of limestone. Even clay contains it in large quantity, both as sand and combined with the alumina which forms that material, and all the slaty rocks and granites are built up of silicates. The old volcanic and igneous rocks, and the products thrown out by modern active volcanoes, consist also largely of silica in combination with alumina, and we are justified in supposing it to be essential to the existence of a world like that we inhabit.

With the greater part of the really important uses of silica we have nothing to do in this chapter. We here regard it only in its crystalline character as a gem, not a gem of great value, and in most cases not rare, but one which, in several ways and with many modifications, is well worthy of careful consideration. There are four groups of valuable quartz minerals used for ornamental purposes, each well marked, and each having its own peculiar interest.

I. Vitreous, or Glassy Quartz.

OBJECTS IN ROCK-CRYSTAL.

The most remarkable mineral of this variety is the clear transparent glass-like stone called Rock-crystal, much used for optical purposes, owing to its perfect transparency, hardness, and refractive power. Being found abundantly in Brazil, and imported thence for lenses and spectacle-glasses, it is known commercially as "Brazilian pebble." It was once believed to consist of petrified ice, as is indicated by its name (Greek, κρύσταλλον, ice), and as is assumed in the following quaint description from Marbodus:—

> *Crystal* is ice through countless ages grown
> (So teach the wise) to hard transparent stone:
> And still the gem retains its native force,
> And holds the cold and colour of its source—
> Yet some deny, and tell of crystal found
> Where never icy winter froze the ground;

94   IN SEARCH OF MINERALS.

> But true it is, that held against the rays
> Of Phœbus it conceives the sudden blaze,
> And kindles tinder, which, from fungus dry
> Beneath its beam, your skilful hands apply.
> Dissolved in honey, let the luscious draught
> By mothers suckling their lov'd charge be quaffed,
> Then from their breasts, as sage physicians show,
> Shall milk abundant in rich torrents flow.

Rock-crystal is the name given to transparent quartz generally, and it occurs both in regular crystalline forms and in fragments and pebbles of all dimensions consisting of water-worn crystals. Though sometimes flawed, it often exhibits perfect transparency. Found in various places in all the British Islands, and in almost every country in the world, it is met with in exceptionally fine crystals in Savoy and Dauphiné. Fine specimens are obtained elsewhere in the old world, in the East Indies and Ceylon, in Madagascar and Tasmania, while large quantities of perfectly transparent stones without a flaw are found in Brazil, as has been already mentioned. The crystal is usually a six-sided prism, terminated with a pyramid; but derived forms, and twin crystals, where the prism is terminated at both ends by similar pyramids, are common. The crystals often contain foreign substances, and not unfrequently blebs, or cavities which are filled with fluid. These cavities have occasionally been met with of very large size, and the fluid extracted on breaking the stone has been examined by the chemist.

The cavities are seldom full of the liquid, there

being generally a bubble or vacant space, the shifting of which attracts attention to the existence of the cavity. It is probable that this is owing to the original formation of the crystal at a much higher temperature than that of the spot where it is now found on the earth's surface. The subsequent contraction of the liquid during cooling has left the vacant space. In one case examined with great care the cavity was only three-fourths full of liquid at ordinary temperatures, but became full at the temperature of $83°$ F., which is less than that of the shade in a hot summer-day. Stones containing cavities filled with fluid were called by ancient writers *Enhydros*, and were regarded as miraculous. The liquid is, however, generally water mixed with saline matter, although in some varieties of quartz very different contents have been met with in the cavities. The following account of such phenomena is interesting and suggestive. It alludes to the notion of crystal being petrified water, already referred to :—

> Streams which a stream in kindred prison chain,
> Which water *were* and water still remain,
> What art hath bound ye, by what wondrous force
> Hath ice to stone congealed the limpid source?
> What heat the captive saves from winter hoar,
> Or what warm zephyr thaws the frozen core?
> Say in what hid recess of inmost earth,
> Prison of fleeting tides, thou hadst thy birth?
> What power thy substance fixed with icy spell,
> Then loosed the prisoner in his crystal cell?[1]

Spheres or balls of crystal are occasionally found

---

[1] Translated from Claudian, Epigram viii. *et seq.*; King, p. 96.

among ancient remains, and are valued highly in all parts of the East, where they are occasionally used to produce results which are thought magical. Pliny describes such spheres as employed in his time to condense the sun's rays for the purpose of cauterising, and Orpheus recommends their employment to kindle sacrificial fires. He also alludes to the flame thus obtained as originating the well-known vestal fires. In recent times a crystal ball held in the hand by an Egyptian boy was said to have enabled him to read future events and see persons and their occupations in places far distant from the place of observation.

Very perfect crystals of quartz found in our own country are known as Cornish diamonds, Irish diamonds, or Bristol diamonds, according to the locality in which they are found, and all are occasionally used for ornamental purposes in their natural state. Large blocks and boulders of transparent rock-crystal have been used for cutting into bowls and other ornaments. Pliny mentions a lady who gave for a crystal vessel a sum

CRYSTAL OF QUARTZ.

equal to £1,500 of our money. A specimen is recorded as existing in the Capitol in Rome weighing fifty pounds, and a bowl is mentioned which held two quarts. A crystal was brought from Delhi, when taken by the English, of an egg-shape, perfectly transparent, and more than 12 inches long. Vast numbers of vessels and ornaments of rock-crystal were brought to England from India to the great Exhibitions of 1851 and 1862, and almost every museum in Europe abounds in specimens of all shapes and sizes.

The name *Iris* is applied by French jewellers to a transparent variety of rock-crystal reflecting the prismatic colours, in consequence of flaws in the interior of the stone. When cut it resembles opal. The same effect may be produced artificially by a blow, or by suddenly dropping it when heated into cold water, or by dropping it when cold into boiling water. The Empress Josephine possessed a remarkably fine set of ornaments of this stone, but the intrinsic value was inconsiderable.

Like other gems, rock-crystal was supposed by the people of the Middle Ages to possess certain occult properties. Thus, when worn by sleepers, it was said to drive away evil dreams, and baffle spells and witchcraft. When powdered and mixed with honey, it was believed to fill the breasts with milk, and thus be valuable to mothers. It is to be feared that no such results would now be obtained.

The *Amethyst* is a variety of rock-crystal, of clear purple or violet-blue colour, probably owing to the presence of a small quantity of manganese, although some chemists have attributed the colour to iron and soda, which are also present in minute proportions. It is a beautiful and often valuable stone. The finest specimens are from various localities in Asia, but good stones are found in different parts of Europe, and a vein of amethyst was opened and worked many years ago at Kerry Head, in Ireland. The following description of the amethyst is tolerably correct, but the gem is now less common than it formerly was, and a vein, if discovered, would not long lie neglected.

> The Tyrian purple the rich *Amethyst* dyes,
> Or darker violet charms the gazer's eyes;
> Bright as the ruby wine another glows,
> Or fainter blush that decks the opening rose;
> Another yet displays a lighter shade,
> Like drops of wine with fountain streams allay'd,
> All these supplied by jewell'd India's mart,
> Easy to cut, yield to the graver's art:
> The gem, if rarer, were a precious prize,
> But now, too common, it neglected lies;
> Fam'd for their power to check the fumes of wine,
> Five different species yields the bounteous mine.

The paler varieties of amethyst are more valued than the more heavily-tinted stones, and were generally selected by the ancients for sculptures of the better class, though scarabei, and other common subjects, are frequently to be met with in dark stones.

The five species of amethyst referred to by the poet, as indicated by the different shades of colour, can be thus described. One was probably the stone sometimes called Oriental Amethyst, and in reality a sapphire; another was the ordinary amethyst, a third includes some darker-tinted stones, the fourth was rose-quartz, and the fifth the citrine.

It has been supposed that the name amethyst is derived from the Greek (α, *neg.*, and μεθύω, *to inebriate*), from the idea expressed in the above quotation, "Famed for their power to check the fumes of wine," but it is more probably derived from the Indian name of the stone. It is certainly of very ancient date.

The value of the amethyst as a gem has varied exceedingly at different times, and according to fashion. A celebrated necklace of these stones, considered to be especially well matched, and belonging to Queen Charlotte of England, was at one time valued at £2,000, but fifty years later the stones would not have fetched in the open market a tenth part of that sum. All that time there had been introduced a large quantity from Germany, and being worked very easily, the stones had become common. They have since become more rare. The violet amethyst is recognised as a fit jewel for persons in mourning, and rings of it are worn by ecclesiastical dignitaries of the Roman Church.

The natural colour of the amethyst can be modified and expelled by roasting in hot ashes, and in this way also shades of colour can readily be obtained by artificial means from dark stones. Many stones vary a good deal in tint in the same specimen.

Rose Quartz is a variety of amethyst used in jewelry. It is transparent, or nearly so, of a delicate rose-pink colour, and usually massive. It is found in Scotland and Ireland, and also in Bavaria and Siberia, generally associated with manganese, from the oxide of which metal it no doubt derives its colour. It takes a fine polish, and when of good colour is sometimes passed off as spinelle, but is inferior in hardness, transparency, and fire to that gem. When long exposed to strong light in a warm place this variety of amethyst loses colour, which is, however, recovered by keeping it for a time in a damp place in the dark.

False Topaz is a light-yellow Brazilian variety of quartz-crystal of some value, and often sold for yellow topaz, from which it differs in crystalline form, hardness, and specific gravity. The colour is rarely free from a smoky tint, which, however, sometimes produces a rich and beautiful effect when not too heavy. It is often cut, and is much valued for seals and brooches. The name of *Citrine* is given to a lemon-yellow, golden, or wine-yellow variety.

Cairngorm, or *Smoky Quartz*, is a variety of rock-crystal dear to the lapidary, and having considerable value as a gem. It is found in the mountain of the same name in Inverness-shire in Scotland, and also in the Grampian Hills. The colour is deep orange or deep brown, and is very beautiful, approaching in some cases the hyacinth, and it is superior in lustre to a similar variety from Germany, called the *German Topaz*. The Cairngorm is frequently manufactured into the handles of dirks and paper-knives, and various articles of Highland costume.

Darker-tinted specimens of rock-crystal, the colour

being charcoal-black or brown-black, are sometimes called by lapidaries *Morion*.

These are the principal varieties of transparent rock-crystal used in jewelry, but there are a few stones of interest that belong to the same group of vitreous or glassy-quartz nearly opaque, owing to the presence of foreign substances caught up during the progress of crystallization. Of these stones *Aventurine* is the most important. It is generally translucent, and consists of scales of mica and occasionally minute flat spangles of metallic copper, generally yellow, bedded in and forming part of the quartz. This stone is found frequently in many parts of Europe and in India; but the finest specimens are from Siberia. The colour is grey, green, brown, or reddish-brown, and it is generally in small pieces, which take a high polish, and are used for rings, ear-rings, shirt-studs, &c. A very large specimen was presented many years ago to Sir R. Murchison by the Emperor of Russia (Nicholas I.), and at his death was presented to the Museum of the Royal School of Mines in Jermyn-street. It is cut into the shape of a vase 4 feet high and 6 feet in circumference. Its colour is pearl grey, with rose tints. Its hardness is considerable, being rather greater than rock-crystal. Only one similar specimen is known, and this was presented to Baron Humboldt, and is in the Museum at Berlin.

Aventurine is better known by imitations than by specimens. The false stone is prepared by heating together to the melting-point of glass pounded glass, protoxide of copper, and oxide of iron, and allowing the mixture to cool very slowly. The name, which signifies chance, is derived from the accident of a workman having let fall brass filings into a pot of melted glass, and thus producing a substance which closely resembles the natural stone.

Venus' Hair Stone is a name given to rock-crystal in which needle-like crystals of titanium are embedded. This curious and beautiful mineral is found in Brazil and Madagascar. The same name is sometimes applied to rock-crystal containing silky tufts of amianthus or asbestos. In a variety found at St. Gothard, crystals of red oxide of titanium cross each other in all directions like the meshes of a net. The variety exhibiting this appearance has been called *Cupid's Net*, or *Love's Meshes*.

Violet quartz containing small golden-brown fibres of oxide of iron is called *Venus'* or *Cupid's Pencils*. Stones of this kind are found in many places in Europe. All these varieties are used by the jeweller.

II. Chalcedonic Quartz.

There is a large group of very beautiful modifications of quartz of which Chalcedony is the type. The stones are not crystals, and differ in this respect from those just described, though like rock-crystal they consist almost entirely of silica. A little water present in the composition is perhaps the cause of the difference.

Chalcedony is so called from the name of a place in Asia Minor, whence it was formerly obtained. It is semi-transparent, hard, and tough. It is some-

QUARTZ GEMS. 103

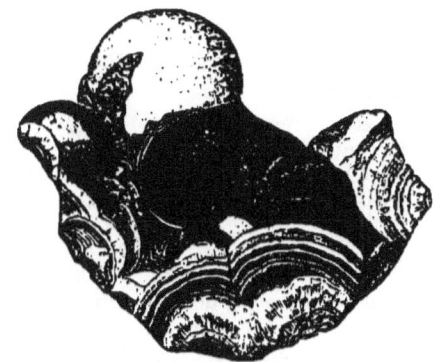

CHALCEDONIC QUARTZ.

times found in rounded lumps almost like grapes, or in parallel bands, so that when broken it presents something of the appearance shown in the above cut. It is generally milk-white in colour, with a more or less bluish tinge. It often occupies cavities in rocks, just as is the case with those kinds of limestone called stalactite and stalagmite. Sponges and other substances of animal origin are often found converted into this stone, and buried in chalk. It is widely distributed, and often under circumstances that indicate volcanic action, whether ancient or modern. It is, however, frequently met with in rocks that have not undergone change of this kind.

Chalcedony was much used for sculpture among the ancients; the earliest Babylonian cylinders, Etruscan scarabæi, and numerous Greek and Roman camei attesting this use. The finest Persian cylinder known, engraved with a king fighting a lion, is formed of this stone. It was no doubt a signet-ring. Referring to this use, its value is thus indicated in the "Lapidarium."

> Between the Hyacinth and Beryl plac'd,
> With lustre fair is the *Calcedon* grac'd ;
> But pierced and worn upon the neck or hand
> A sure success in lawsuits 'twill command.

Prase or *Plasma* (from the Greek πρασὸν, a leek), is a faintly translucent chalcedony, of leek-green colour, having a glistening or waxy lustre, interspersed with spots or patches of dull yellow, white, or black. It sometimes approaches the emerald in colour, but is devoid of lustre. As a stone for sculpture it was much used at one time, but never highly valued, and does not seem to have been employed by the best artists. It is spoken of by Pliny very contemptuously in the sentence "Vilioris est turbæ Prasius" (the Prase belongs to the vulgar herd). It has, however, been often met with in camei or intagli among the *débris* of the ancient buildings of Rome :—

> Midst precious stones a place the Prase may claim,
> Of value small, content with beauty's fame,
> No virtue has it ; but it brightly gleams
> With emerald green, and well the gold beseems ;
> Or blood-red spots diversify its green,
> Or crossed with three white lines its face is seen.

The *Chrysoprase* is an opaque apple-green variety of the Prase, and, like it, is a chalcedony having little lustre, but capable of being cut into many ornamental shapes. It is hard, and comparatively rare. At present it is found in Silesia, and, rarely, in North

America. According to Marbodus it once came from Africa :—

> In Afric springs the Chrysoprasion bright,
> Which day conceals, but darkness brings to light :
> By night a shining fire, it lifeless lies
> Like golden ore when day illumes the skies.
> Revers'd is Nature's law where light reveals
> Whate'er in darkness shrouding night conceals.

The colour of Chrysoprase is probably derived from oxide of nickel, of which it contains about one part in a hundred. The reference in the above lines to one peculiarity of this stone is very curious. It loses colour and becomes dark and clouded if kept in a dry, warm situation, or exposed to strong sun-light; but the colour may be restored if the stone is kept in a damp place, or dipped in a solution of nitrate of nickel. This latter treatment improves inferior stones.

Though not of great value, the rarity of good stones, which were formerly monopolized by the Prussian authorities, and the fact that a necklace of them is considered a charm against pains and aches by Silesian peasants, has given a factitious value to this gem.

> Parent of gems, rich India from her mines
> The *Chrysoprase*, a precious gift, consigns,
> As leaves of leeks in mingled shadow blent,
> Or purple dark with golden stars besprent ;
> But what its virtue, rests concealed in night :
> All things Fate grants not unto mortal sight.

Of the chalcedonic varieties of quartz the *Carnelian*

is one that is often much admired. It is of a clear, bright red or flesh-coloured tint, whence the name (Latin, *carnis*, flesh), and the gradation from red to white passes curiously, and often very beautifully, through flesh-red and blood-red, varied with brown, orange-coloured and yellow tints. The best and most valued specimens, however, are of uniform colour throughout, and free from the cloudiness and muddiness generally met with in inferior stones.

The following curious and quaint account of this stone is quoted from Philemon Holland's racy translation of Pliny's treatise on this department of natural history:—

"In old time there was not a pretious stone in greater request than the *Cornalline*, and, in truth, Menander and Philemon have named this stone in their Comœdies, for a brave and proud gem: neither can we find a pretious stone that maintaineth the lustre longer than it, against any humour wherein it is drenched; and yet oile is more contrarie unto it than any other liquor. To conclude, those that be of the colour of honey are rejected for nought; howbeit, if they resemble the colour of earthen pots, they be worse than those."—*Pliny*, Book xxxvii. ch. 7.

The properties assumed to belong to the Carnelian are expressed in the following lines from the "Lapidarium:"—

> Let not the Muse the dull Carnelian slight
> Although it shine with but a feeble light;
> Fate has with virtues great its nature grac'd.
> Tied round the neck or on the finger plac'd,

> Its friendly influence checks the rising fray,
> And chases spites and quarrels far away :
> That, where the colour of raw flesh is found,
> Will stanch the blood fast issuing from the wound ;
> Whether from mangled limbs the torrents flow,
> Or inward issues source of deadly woe.

The *Sard* is a valuable variety of Carnelian of more even quality, superior hardness and toughness, and higher colour than the ordinary stones bearing the name. It retains a high polish, and is supposed in this respect to excel other quartz gems, and even some of the hard group, many ancient sculptured gems of sard being perfect, while harder gems, as garnets and hyacinths, have been scratched and roughened by wear and exposure. The bright red varieties are the most valued, and they often approach the carbuncle in lustre. The name Sard is derived from Sardis, whence the stones were first imported into Greece, but it is probable that the best specimens always came from the interior of Asia. At one time they were common in ancient Babylon, but afterwards they came from India and Arabia, being found on the shores of the Red Sea. The sard, though not extremely valuable, is rather unfairly described by the old poet, Marbodus, when he speaks of it as

> Cheapest of gems, it may no share of fame
> For any virtue save its beauty claim :
> Except for power the Onyx' spell to break.

The *Onyx* is a very curious variety of chalcedonic quartz, consisting of two, or sometimes more, opaque layers of different colours, generally in strong contrast, as black and white, dark red and white, green and white, and many other varieties. When there are three layers, the uppermost is either red, blue or brown, and under this is a stratum of white, the lowest bed being jet black or deep brown. These conditions are very favourable to the ingenuity of the sculptor, who is able to take advantage of the colours to produce very singular effects. Like some other conditions in minerals these varieties of colour are not seldom produced artificially by the jeweller, stones, otherwise of little value, being boiled for a long time in honey-and-water and then soaked in sulphuric acid to bring out the black and white, and in nitric acid to give red and white layers.

> Called by the *Onyx*, round the sleeper stand
> Black dreams, and phantoms rise, a grisly band :
> Whoso on neck or hand this stone displays,
> Is plagued with lawsuits and with civil frays ;
> Round infants' necks if tied, so nurses show,
> Their tender mouths with slaver overflow.
> This the Arabian, this the Indian sends,
> And five the sorts to which its name it lends :
> Which name of Onyx, as grammarians teach,
> Comes from the usage of the Grecian speech,
> For what the name of *nails* amongst us bears,
> Expressed in Greek as *Onyches* appears.
> Yet if a Sardian on thy finger shine,
> 'Twill quash the Onyx' influence malign.

The onyx is not a common stone. It was found by the ancients in Arabia, and possibly was obtained also from India and the steppes of Tartary. Specimens have been met with in Scotland, in Perthshire, and in the Isle of Skye, and in Ireland, at the Giant's Causeway, and, curiously enough, on the shores of Lough Neagh. The onyx was one of the stones ordered to be worn on the garments of the High Priest of the Jews, and was highly regarded in many ways by Eastern people.

Onicolo is a name given to a variety of the onyx, consisting of a deep brown or black ground over which is a band of bluish tint. The word is the diminutive of onyx, but, as it is sometimes changed to *Nicolo*, it has been supposed by some to be derived from the name of an artist. It was greatly used for fine Roman camei, for which it is well adapted. It is generally of very fine texture.

The *Sardonyx* is beyond comparison the most valuable of the varieties of chalcedonic quartz, and is composed of alternate layers of sard and nearly opaque white chalcedony. It is a very beautiful stone, and is rare. By the ancients it was selected for some of the finest camei, and in many cases was treated with the most consummate ingenuity and art.

> The Sard and Onyx in one name unite,
> And from their union spring three colours bright.
> O'er jetty black the brilliant white is spread,
> And o'er the white diffused a fiery red:
> If clear the colours, if distinct the line,
> When still unmixed the various layers join,
> Such we for beauty and for value prize,
> Rarest of all that teeming earth supplies.
> Chief amongst signets it will best convey
> The stamp impressed, nor tear the wax away.

> The man of humble heart and modest face,
> And purest soul the *Sardonyx* should grace;
> A worthy gem, yet boasts no mystic powers:
> 'Tis sent from Indian and Arabian shores.

Scipio Africanus the Elder is said to have first made the sardonyx fashionable in Rome, and it was also worn by the Emperor Claudius. It is described by Pliny as a white opaque layer superimposed on a red transparent stratum of true sard, and this description applies exactly to the Indian stones of this kind. The Arabian specimens were different, the lower stratum being black or blue, covered with opaque white and an upper plate of red. The Indian were the most valued, and were often pierced to be worn as necklaces.

Sardonyx has been imitated, both anciently and in modern times, by placing a sard on a red-hot iron, thus converting the surface into an opaque white layer, which forms a good relief when cut through in making a cameo to a red layer below.

Very beautiful and valuable camei cut in sardonyx are described. One of extreme beauty, measuring $7\frac{1}{2}$ inches by 6 inches, was sold in 1859 for £126, and another, little more than $1\frac{1}{2}$ inches across, for £31, at the same sale. Very large stones, much larger than the former of these, were used by the ancient engravers—so large, indeed, that it has been thought they were artificially produced by a white cement on a sard.

Agate.—The Agate and the Onyx differ but little in reality, though the appearance is often such as to indicate a great divergence. The layers which in the onyx are parallel, and give the appearance of regular strata, in the agate are sometimes concentric, frequently angular, and sometimes wavy. The colours also differ, and the stones are generally less valuable.

> Achates stream, which through Sicilia's plains
> Winds his soft course, renowned in pastoral strains,
> Named from himself the *Agate* first disclosed—
> A jet black stone by milky zones inclosed :
> With figured veins its various surface strew'd,
> Painted by nature in a sportive mood.
> Now regal shapes, now gods its face adorn ;
> Such the famed Agate by King Pyrrhus worn,
> Whose level surface the nine Muses grac'd,
> Round Phœbus with his lyre in order plac'd.
> Strange to relate, 'twas to no artist due,
> Nature herself the wondrous picture drew.
> Another Agate yields the Cretan shore,
> As coral red, with gold-dust sprinkled o'er ;
> An antidote against the poison'd draught,
> And for the treach'rous viper's venom'd shaft.
> Whilst on that Agate which dark Indians praise
> The woods arise, the sylvan monster strays :
> Placed in the mouth 'twill raging thirst appease,
> And its mild radiance the tir'd eyeballs ease.
> One fumes like myrrh if on the altar strew'd ;
> Another is besprent with drops of blood :
> Whilst those, which like the comb with yellow gleam,
> Are most abundant, but in least esteem.
> The Agate on the wearer strength bestows,
> With ruddy health his fresh complexion glows ;
> Both eloquence and grace by it are given,
> He gains the favour both of earth and heaven :
> Anchises' son, by this attendant saved,
> O'ercomes all labours, every danger brav'd.

Agates are easily distinguished from other stones, their colours being arranged in colour-spots, or bands, but they comprise many varieties of colour as alluded to in the above lines. They are very abundant in Brazil, being brought down from the interior and rolled along the bed of the great River de la Plata. They are thence put on board ships as ballast and thus brought to Europe, where they are cut and polished. Near Oberstein, a picturesque village in the narrow valley of the Nahe, a tributary of the Rhine, shut in by romantic cliffs and entering the main stream near the town of Bingen, many agates were formerly found, and much agate-cutting is carried on. At Galgernberg, in the North of Germany, in Upper Egypt, in Siberia, in Arabia and India, and near Perth, in Scotland, stones of this kind are very common, and have been known almost from time immemorial. To become valuable, however, they must be cut and polished, and this work, though carried on in most parts of the world to some extent, is most cheaply effected at Oberstein. The polishing is effected by water-power obtained from the river and from numerous small streams adjoining the agate quarries. The grindstone used is a coarse red sandstone of the neighbourhood.

The banded agates are the most admired. These bands are sometimes parallel and horizontal, but more often curved or zig-zag: *Ribbon agate* is the name given to the former, *Fortification agate* to the latter.

The colours are often very beautiful naturally, and they are rendered more varied, brighter, and deeper, by boiling the stone in oil and afterwards exposing it to the action of sulphuric acid. The oil penetrates between the layers of which the banded stones are made up, and the acid, by decomposing the oil, both modifies and improves the colour.

Agates are worked up into many forms, and are sometimes employed for useful as well as ornamental purposes. They are found very useful for burnishing and for making into mortars for chemical purposes, owing to their hardness and indestructibility. As seals, brooches, beads, and handles of daggers and knives, they are well known. As cups and vases they are more rare. Occasionally fine specimens are cut into complete services or into tables. Magnificent examples of this kind have been displayed in those great international collections, in which, since 1851, all curious and costly works of art have been exhibited. It is said that a service of this kind in the Treasury of France is valued at half a million francs (£20,000).

Agates are sometimes found in quarries, and most commonly in rocks of igneous origin (basalts, porphyries, or amygdaloid), and this is the case on the Rhine as well as near Perth, both well-known localities.

Mocha-stone, or *Moss Agate*, is a pretty variety of Chalcedony, in which are moss-like markings of various shades, and brown marks resembling trees and vegetable filaments, all probably caused by the infiltration of manganese or iron oxide. Very beautiful specimens have been found in Arabia, whence the name

Mocha-stone, and also in parts of India. Sometimes the stone presents the appearance of a landscape, trees and water being apparently delineated by nature on the flat surface exposed. Such stones are called *Landscape Stones*, but they must not be confounded with the Cotham marble, which is an argillaceous limestone.

The *Cat's-eye* is the last of the varieties of the chalcedonic quartz. It is chiefly found in Ceylon and India, but also in Bohemia and Bavaria. It is a curious stone, more or less translucent, and sometimes quite transparent. It is always massive, and occurs in rounded pebbles seldom larger than a hazel-nut and generally much smaller. Its colour is yellowish, greenish, grey, or yellowish-brown, but it is also occasionally brown, olive-green, red, or even black. It has a peculiar shining lustre by which it is recognised, and when cut *en cabochon*, or so as to present almost a hemisphere, displays a peculiar opalescent, or floating lustre, like the pupil of a cat's eye. This is supposed to be caused by the presence of small fibres of asbestos. The value of the stone depends chiefly on the play of colour.

III. Jaspery Quartz.

Jasper is a perfectly opaque variety of quartz of dull red, yellow, brown, or green colour, generally due to the presence of iron oxide.

> Bright are the Jasper's tints, with clouds,
> And spots, and diverse stripes and splendid veins
> Of green and various hues ; in mass opaque,
> But in thin fragments pervious to the light ·

> With earthy fracture angularly sharp,
> Less hard than flint, but striking fire with steel.
> Jasper in large elliptic masses oft
> Occurs, or nodes detached, or rocks entire,
> To which Egyptian pebble's near allied.[1]

The use of jasper by the ancients as a seal-stone and for purposes of sculpture was very considerable, and all the principal varieties of colour are recognised. The black, a very hard and very fine material, and a dark green hard variety were especially valued, and sculptured examples of these and of a dull yellow variety were valued as talismans. The red jaspers were generally softer and less appreciated; one of them, however, of a very rich crimson, was greatly valued and is very rare. The following lines allude to the varieties recognised, and point out the supposed virtues the stone possesses and the mode in which its beauties are best shown. Like other quotations of the same kind it is a translation of part of the "Lapidarium."

> Of seventeen species can the *Jasper* boast
> Of differing colours, in itself a host.
> In various regions is this substance seen:
> The best of all the bright translucent green;
> The greatest virtue is to this assign'd;
> Fevers and dropsies feel its influence kind.
> Hung round the neck it eases travail's throes,
> And guards the wearer from approaching woes.
> Power, too, it gives when blest by magic rite:
> And drives away the phantoms of the night;
> But let the gem enchas'd in silver shine,
> And fortify thereby its force divine.

[1] "Werneria, or Short Characters of Earths," by Terræ Filius, 1805. Pp. 78-79.

The varieties of jasper here alluded to present only accidental and unimportant differences of colour. Pliny says that the best sort had a tinge of purple, the second of rose-colour, the third of emerald, and that another was also like an emerald but with a white line passing through its middle. This last variety was called by a special name, *Grammatias*, and was used in the East as an amulet. A fifth kind was called by the Greeks *Borea*, and is described as resembling the colour of the sky on an autumnal morning. Those only who are familiar with Greece and its atmosphere at this delightful season can realize the beautiful blue of the tint. It could hardly have applied to any of the stones that would now be called jasper. The other varieties are not now known by special names.

Bloodstone or *Heliotrope* is the name given to a deep green jasper interspersed with deep blood-red spots. Its colour is very beautiful, and the stone is hard and valued for seals, rings, and other ornamental purposes. It takes a high polish. The stones that are most nearly translucent and have the largest number of red spots are the most highly valued. The name *heliotrope* was given from two Greek words (ἥλιο-, *the sun*, and τρέπω, *I turn*), intimating that it turned the image of the sun to blood when reflected from it. The following quotation explains sufficiently the mediæval notions on this subject, and requires no further remark :—

> The Heliotrope or "gem that turns the sun,"
> From its strange power the name has justly won :
> For set in water opposite his rays
> As red as blood 'twill turn bright Phœbus' blaze.

QUARTZ GEMS.

And, far diffused the unauspicious light,
With strange eclipse the startled world affright.
Then boils the vase, urged by its magic power,
And casts far o'er the brim the sudden shower;
As when the gloomy air to rain gives way
It storms evokes, and clouds the fairest day;
It gifts the wearer with prophetic eye
Into the Future's darkest depths to spy.
A good report 'twill give and endless praise,
And crown thy honour'd course with length of days.
It checks the flow of blood, the wearer's soul
Shall laugh at treason or the poison'd bowl.
Though with such potent virtues grac'd by heaven,
One yet more wondrous to the gem is given.
This with the herb that bears its name unite
With incantation due and secret rite,
Then shalt thou mortal eyes in darkness shroud,
And walk invisible amidst the crowd.
The stone for colour might an emerald seem,
But drops of blood diversify the green.
'Tis sent sometimes from Ethiopia's land,
Sometime from Afric or the Cyprian strand.

A velvet-black flinty slate found originally in the province of Lydia in Asia Minor, and thence called *Lydian-stone*, is a kind of jasper, not recognised indeed as a gem, but much valued owing to its smoothness and hardness as a test-stone for gold and silver, the colour of the streak left on the stone when the metals are rubbed on it enabling an experienced eye to detect the amount of alloy. The ancients who used this stone called it sometimes *Lapis Basanites*, and in mineralogy it is called *Basanite*.

Occasionally black flints are found of very curious and grotesque forms, often the result of some coral or shell on which they have been formed and sometimes merely from accident. The interiors of such flints when broken across also sometimes put on strange appearances. One offered a singular resemblance to a tiger and a boar's head facing each other. Other quaint resemblances have often been taken for fossil birds or animals. It is true that the flints in the chalk have generally been formed on some substance that once belonged to a living animal, but except where the animal was that living in a shell, or depositing coralline or other calcareous substance, the fanciful resemblance has no origin in reality.

IV. Opalescent Quartz.

The very beautiful and valuable gem called *Opal* is another variety of quartz, although much less hard, and never found crystalline. It is semi-transparent or pellucid, and very brittle. The colour is milk-white or grey varying to brown, its lustre often pearly or resinous. It is, however, most remarkable for the brilliant play of colours especially characteristic of the valuable stones of this kind, which are thence called Noble or Precious Opal. This play of colours is not very clearly accounted for; Dr. Brewster attributed it to the presence of internal fissures and cracks, and Haüy, a very eminent German mineralogist, ascribes it to thin films of air in cavities in the interior of the stone. Opal consists only of silica with five or ten per cent. of water and a small quantity of alkaline earths. The water is considered to be present mechanically and varies much in different specimens.

Opal came to the Romans from Italy, but in

recent times has chiefly been obtained from Hungary. It is also sent from Honduras, where there are important mines of it. The ancient stones were all small, none exceeding the size of a hazel-nut, this being the size of a celebrated gem valued at £20,000 of our money. It is recorded of a fine opal that its possessor, Nonius, preferred going into exile rather than authorize the forced sale of a ring in which it was set to Mark Antony.[1] It is still very highly esteemed, but the mines of Hungary have yielded some of much larger size, now serving as ornaments to the Austro-Hungarian crown.

The opal requires great care in use as an ornament. It sometimes cracks on being held near a fire in cold weather, and long wear and exposure will close up with dust and grease the innumerable cracks on which are believed to depend its beautiful iridescence and fire. It is said, indeed, that by roasting an opal in this state its former beauty can be recovered, but the experiment involves risk. Transparent opal immersed in melted white wax becomes opaque.

The term *Precious Opal* includes stones which exhibit a rich play of prismatic colours. Such stones are rare and very valuable, moderate sized specimens of good quality having often fetched the price of diamonds of the same size. The largest known specimen is in the imperial museum in Vienna. It is 4¾ in. long, 2½ in. thick, and weighs 17 ounces. It

[1] The ring in which this stone was set had been valued at 20,000 sesterces (£177,083).

has been valued at £70,000. Good opals, when well set, are often surrounded by diamonds. The finest specimens are sometimes called *Oriental Opals*, but this is merely in accordance with the feeling that all the finest gems come from the East.

The finest opals are not unfrequently found in porphyry, of which, indeed, this variety of quartz sometimes forms a component part, and very beautiful ornaments (snuff-boxes and other objects) can be made out of the matrix including the opal, thus presenting the stone very favourably, but the greatest care is needed for this, as the opal is both soft and brittle.

When held between the eye and the light a fine opal appears of a pale red and wine-yellow tint with milky transparency. By reflected light, as its position is altered, it displays the most beautiful play of colours, particularly pale green and emerald green, golden-yellow, fire-red, bright blue, rich violet, purple and pale grey. When all these colours are displayed in the same specimen, arranged in small spangles, it is called *Harlequin Opal*. Sometimes only one colour is present, and in this case, when the colour is rich orange-yellow, it is called *Golden Opal*. Vivid emerald green tints are also extremely beautiful, but have no special name.

A milk-white transparent variety of opal, reflecting a reddish colour when turned to the sun or any bright light, is called *Girasol* or *Fire-opal*. Sometimes this variety resembles jelly in its reflection of light.

Another curious variety, readily absorbing water, and, though not naturally transparent, becoming so when immersed in water, is called *Hydrophane*, and by the ancients *Oculus Mundi* (eye of the world). A dull variety of bluish-white colour and showing a pearly lustre beneath the surface is called *Cacholong*. Hydrophane immersed in white wax, when perfectly dry, absorbs a certain quantity and becomes translucent or even transparent.

A transparent or semi-transparent opal found in small lumps in some volcanic rocks, and nearly resembling clear glass, is called *Hyalite* or *Müller's Glass*. It is found in Hungary, the country of opal, and also in Bohemia, and has been met with in England and Ireland.

There are several other varieties of opal, but they are not used for ornamental purposes, and cannot therefore be included in the present chapter. They are all of the same composition, containing more or less water, and occasionally a small percentage of foreign substances.

The following curious account of the opal is quoted by Mr. King, in his "Antique Gems," p. 422. It is by a certain Petrus Arlensis, writing in 1610. It is not often that the contemplation of a mineral has so excited the imagination:—

"The various colours in the opal tend greatly to the delectation of the sight, nay, more, they have the very greatest efficacy in cheering the heart and the inward parts, and specially rejoice the eyes of the beholders. One in particular came into my hands, in which such beauty, loveliness, and grace shone forth, that it could truly boast that it forcibly drew all other gems to itself, while it surprised, astonished, and held captive, without escape or intermission, the hearts of all who beheld it. It was of the size of a filbert, and clasped in the claws of a golden eagle wrought with wonderful art, and had such vivid and varied colours that all the beauty of the heavens might be viewed within it. Grace went out from it, majesty shot forth from its almost divine splendour. It sent forth such bright and piercing rays that it struck terror into all beholders. In a word, it bestowed upon the wearer the qualities granted by Nature to itself, for by an invisible dart, it penetrated the souls and dazzled the eyes of all who saw it; appalled all hearts, however bold and courageous; in fine, it filled with trembling the bodies of the bystanders, and forced them, by a fatal impulse, to love, honour, and worship it. I have seen, I have felt, I call God to witness, of a truth such a stone is to be valued at an inestimable amount."

Semi-opal is a dull variety of opal differing from the finer kinds by its greater opacity, and like common opal it does not exhibit in full beauty the delicate play of colours so remarkable in the gem. It is, however, often used for ornamental purposes, such as pins and cane-heads. The name *Pyrophane* is given to a variety which, when heated in a spoon, becomes transparent, but returns to its opaque state when cool. It has been said that some stones, found in Armenia, are rendered transparent by exposure to a hot sun, but change at night, and become nearly opaque when the sun has set.

Tabasheer is a concretion formed in the interior of the stem of the large Indian bamboo, bearing some resemblance to Hydrophane and consisting of silica. It is imperfectly transparent and often opaque, but in any case when dipped in water it gives out a quantity of air-bubbles and becomes more transparent, returning to its original state when dried. It is very light when dry, and its weight is doubled when soaked; it therefore absorbs a large quantity of water. In the East, Tabasheer is regarded as a valuable medicine, and spoken of by names which mean bamboo milk, bamboo camphor, and bamboo salt.

CHAPTER V.

The Softer Gems and Valuable Stones.

We have regarded quartz as forming a complete intermediate class between the hard gems, which are generally those of greatest money value, and the softer gems tnat can be scratched by quartz, and must only be used with care, and with attention to their texture, which renders them liable to injury by exposure or carelessness. The number of the stones of this kind is large, and some of them have considerable value, though as a whole they are not regarded as precious. They differ greatly in composition, some being compounds of only two elements, and some offering extreme complication. A few are found and used in the crystalline state, but most of them, and the most valuable, are not so. Many of them are stones not showing crystalline texture.

Chrysolite.

This stone is regarded as a gem, and is known by various names. It is a compound of silica and magnesia, a certain portion of iron replacing some of the magnesia. It is brittle and not very hard, and is not altogether fit to be manufactured into personal ornaments. It is also deficient in play of colour. Its usual colour is greenish-yellow, and its lustre brilliantly vitreous. When the stones are large, of good tint, well matched, free from flaws, and well cut and polished they are made into necklaces and other ornaments with some effect, but as single stones they are rarely of great importance. The name *Oriental Chrysolite* is sometimes given to good specimens. The stones are more generally found as pebbles than as complete crystals. They occur in Upper Egypt and near Constantinople, and among volcanic rocks in Auvergne, near Vesuvius, in the Isle of Bourbon, and in Mexico.

The chrysolite is also known as *Peridote* and *Olivine*, the difference being unessential, though olivine is not used for ornamental purposes. The modern chrysolite is the topaz of the ancients, and, notwithstanding its softness, it was by them much valued, and even used for sculpture. The yellow varieties only were called chrysolite, and these also are the harder and more valuable. By artificial light, indeed, they possess almost the lustre of the diamond, often appearing of the purest water when the colour is not seen by daylight. In the peridote green is the prevailing colour, and the stone often resembles a rolled pebble of bottle-glass.

> The golden *Chrysolite* a fiery blaze,
> Mixed with the hue of ocean's green displays;
> Enchased in gold its strong protective might
> Drives far away the terrors of the night :
> Strung on the hairs plucked from an ass's tail,
> The mightiest demons 'neath its influence quail.
> This potent amulet, of old renown'd,
> Wear, like a bracelet, on thy left arm bound.
> 'Tis brought by merchants from those far-off lands
> Where Ethiopia spreads her burning sands.

The chrysolite and peridote have been sometimes confounded with the chrysoberyl, the greenish-yellow varieties of sapphire, the topaz, the aquamarine, and even apatite, idocrase, and green tourmaline. It is softer than the three former, but harder and heavier than apatite. It differs from tourmaline in having no electrical properties.

Idocrase.

This stone is sometimes cut into ring-stones, and other simple ornaments sold at Naples and Turin, and it often passes as chrysolite. It is generally found crystalline, and is most commonly obtained from Piedmont. Some of the crystals are clear, and of fine pale-green colour. It consists of a silicate of alumina and lime, and has little value as a gem.

Turquoise.

This singular stone is a phosphate of alumina coloured by oxide of copper, and is obtained chiefly from Persia and Arabia, where it appears to be common in certain limited localities. One of these is in Khorassan, the mines of which are referred to by Tavernier, and there are also several mines about thirteen days south-east of Suez. Good specimens have been obtained from Abyssinia, and inferior stones are known elsewhere.

The turquoise is often rounded, or kidney-shaped, and resembling an incrustation; it is feebly translucent, nearly opaque, and of a peculiar bluish-green colour, which is very constant.[1] It is never crystalline. The hardness is nearly equal to that of the agate. Nodules of turquoise are often found in groups, like currant seeds in sandstone.

Among the peculiarities of this stone is that when kept near musk or camphor, or buried in damp earth, or exposed to fire, the colour fades, and it is even said to be the case that clearness or dulness of tint is produced by mere changes in atmospheric condition. The colour and surface are certainly injured when the stone is exposed to light and moisture.

The ancient *Callais* is the same stone as that we now call turquoise. It is described as a stone " which grew upon its native rock in shape like an eye, was cut, not ground into shape, set off gold better than any other gem, was spoilt by wetting with oil, grease, or wine, and was the easiest of all to imitate in glass." It was the favourite ornament of the ancient Caramanians, and is still preferred by the modern Persians, who lavish it in profusion over all their trappings and weapons. There was a mediæval notion that it grew pale on the finger of a sickly person, but recovered its colour on removal to a healthy hand. Another fancy was that its hue varied with the hour of the day, so that to the careful observer it could serve the purpose of a dial. In Germany it was believed that when presented as a love gift its colour would remain

---

[1] It would perhaps be more correct to say, that the same mineral having other colours, or being colourless, has no value, and is therefore not brought into the market.

unaltered so long as the giver was faithful, but would grow pale if his affection should fade.

"The turquoise," we are told, "is good for liberty, for he that hath consecrated it, and duly performed all things necessary to be done in it, shall obtain liberty. It is fitting to perfect the stone when you have got it in this manner: engrave upon it a beetle, then a man standing under it; afterwards let it be bored through its length, and set on a gold fibula (swivel); then being blest and set in an adorned and prepared place, it will show forth the glory which God hath given it."

We are also told by mediæval authority that turquoise is useful for riders. "As long as one wears it his horse will not tire, nor throw him. It is also good for the eyes, and averts accidents."

A large and fine set of these gems was brought over from Arabia by Major M'Donald about thirty years ago, and placed in the Great Exhibition of 1851. Many of the specimens are at the Museum of Practical Geology. Besides the stones in various conditions, and some contained in the sand rock, which is the matrix, there are several veins of turquoise from a twentieth to a tenth of an inch thick, which cut across the bedding of the sandstone like fine threads.

Owing to the perishable nature of the material, sculptured specimens prepared by the ancients have not come down to us; but in addition to the true, or as it is sometimes called the Oriental turquoise, there is another inferior kind called *Odontolite*, *Bone Tur-*

quoise, or *Occidental Turquoise.* This stone is able to withstand the heat of a fire, and is frequently used for ornament. It is an artificial gem, and appears to be a kind of petrified bone. It is considerably less hard than the other, and was much used for sculpture. Some works executed in this material are of great beauty.

Lapis Lazuli, or Azure Stone.

This very beautiful stone, owing to the extreme richness and beauty of its colour, is frequently polished and used for certain kinds of jewelry, besides being cut into vases and other ornaments and employed in Mosaic work. It is a silicate of alumina, lime, and soda, but it also contains sulphides of iron and sodium. It is the material used in the composition of the exquisite and costly colour called ultramarine. It is nearly opaque, or only translucent at the edges, and never crystalline. It is chiefly brought from Siberia, Persia, and China, and its home is described by Mr. Prinsep[1] in the following words:—

"The country of Badakshan abounds in mountains, and contains several rivers. On the Oxus river, near where the Samarkand road crosses it, is the mine of lapis lazuli. This mineral has different shapes; one like the egg of a hen, which is covered with a thin, soft, and white stony coal, is reckoned the best when pounded; it needs neither washing nor polishing; the others are without covering, and must be

"Oriental Account of Precious Minerals."

washed. The method of washing is this:—first to pulverize it, and afterwards to keep it wrapt in silk cloth, besmeared all over with gum sandarach, which should be previously softened in very hot water, and then rubbed over or kneaded with the hands; it is kept in the water for three days, until all the foreign matter has been washed out."

Lapis lazuli is the mineral described by Pliny under the name of sapphire, when he states "that it came from Media" (whence the principal supply was brought till quite recently), "that it was opaque, and sprinkled with specks of gold" (iron pyrites), "and that it was of two sorts, a dark and light blue." It was considered unfit for engraving upon in consequence of its substance being full of hard points (pyrites). Notwithstanding this, there are many very fine works of art sculptured in it.

Besides sapphire, the name of *Cyanos* has been applied to lapis lazuli, probably to certain earthy varieties that were used to manufacture colour. Some of the accounts would rather indicate that a cobalt mineral was alluded to under this name. It is often difficult to identify minerals as spoken of under various names, varying according to accidents of form and colour. In some cases there can be no doubt, but in others much careful consideration is necessary before a satisfactory result can be arrived at.

Malachite.

Among the valuable stones which from their use for ornamental purposes may be regarded as precious we must not exclude the beautiful green carbonate of copper known as Malachite. This mineral occurs but rarely crystalline, and the crystal, though not without beauty, is soft, and not used by the lapidary. The usual appearance of the stone is massive, in globular or kidney-shaped lumps, and of concentric structure, frequently fibrous and silky. It is strictly a copper stalactite or stalagmite, corresponding with the limestones thus named in being formed by the evaporation of water, containing the mineral in solution, which has dropped from the roof of a cavern, and on evaporating has left the carbonate of copper behind in thin films successively added. Owing to the mode in which it has formed, the curved surfaces in the stone appear to intersect, and the lines marking the deposit seem broken. As the lines are very clearly marked, and almost every stratum has some slight peculiarity of tint, the variety is extremely great. The stone is capable of receiving a very high polish, and, being moderately soft, is easily cut. The usual colour is the pale green of the leaves of the marsh-mallow, whence the name, derived from the Greek, μαλακή, or marsh-mallow, but there are infinite varieties of shade, and occasionally the green passes into blue. There is a blue variety of the carbonate of copper, however, known by a different name, *Azurite*, or *Chessylite*, but though less common, and equally beautiful, this mineral is rarely used for ornamental work. It is generally found crystalline.

Magnificent specimens of malachite have been found in the Siberian copper-mines, where, on one occasion, a mass 18 feet long and 9 feet wide was cut into. It was estimated to contain 500,000 lbs. (nearly 250 tons), of pure and solid malachite, and its value was enormous. From it were constructed, by a very ingenious process of inlaying and veneering on an iron surface, a pair of the most magnificent doors ever produced. These, with inlaid tables, vases, and other articles of furniture, formed some of the most striking ornaments of the Great Exhibition of 1851. They were sent from Russia, and exhibited by Prince Demidoff. They are represented in the Frontispiece. Noble specimens of the mineral, but not so large as this or of such quality, have been found since in the well-known copper mines of Burra Burra, in Australia, but these are not generally so well-adapted for inlaid-work, and involve greater loss of material, when used for ornamental purposes. Small specimens of malachite are common in Cornwall and elsewhere. They are, indeed, met with in almost every copper-producing district, but these also are less fitted for inlaying than the best Russian specimens.

It is supposed that malachite was the *Chrysocolla* of the ancients. It is said that Nero, in one of his extravagant fits, caused the floor of the circus to be strewed with the powder of this valuable and costly mineral instead of with sand as was usual. The *Molochites* of classical authors is also supposed to have been malachite. Pliny, writing concerning it, says, quoting the words of his quaint translator, Dr. Philemon Holland, "Commended it is highly in signets for to seale faire ; and besides it is supposed to be, by a natural virtue, that it hath a counter charme to preserve little babes and infants from all witchcrafts and sorceries" (Bk. xxxvii. cap. 8).

There are a few, and only a few, antique camei in malachite. Among them is one of the best period of Roman art. Its beauty, however, can hardly be judged of, as it is coated with brown oxide, with which it was entirely encrusted at the time of its discovery.

The supposed qualities of malachite are thus expressed by Marbodus :—

> The *Molochites'* virtue keeps from hurt
> The infant's cradle all mischance to avert,
> Lest spiteful witchcraft blast the tender fame,
> Virtue with beauty joined exalt its fame.
> Opaque of hue, with th' Emerald's vivid green,
> It charms the sight, first in Arabia seen.

Labradorite or Labrador Felspar.

This very beautiful material is a variety of felspar extremely remarkable on account of its rich chatoyant reflections. It is a silicate of alumina, lime, and soda, chiefly met with as a constituent of rocks, but often occurring in blue crystals. It was originally noticed on the shores of Labrador, whence its name, but has been found in many parts of Europe, and in the United States generally, in porphyritic rocks of various kinds.

" Besides the fundamental colour, this mineral, which takes a very fine polish, presents a most beautiful play of vivid tints, varying according to the posi tion in which it is viewed. Of blue, it exhibits all the varieties from violet to smalt blue; of green, it displays the pure emerald-green, and various other tints, approaching to blue on the one hand and to yellow on the other. Of yellow, the most usual shades are golden and lemon yellow, verging into deep orange, and thence into rich copper-red and tombac-brown. The parts exhibiting these colours are disposed in irregular spots and patches, and the same spot, if held in different positions, displays various tints; of these violet and red are the most rare." The singular play of colour is not very clearly explained. It is, to some extent, common to all the varieties of felspar, but is in none else so striking.

Good specimens of Labradorite are manufactured into brooches, pins, bracelets, &c., but it is more frequently used for snuff-boxes. It requires great skill in cutting, to show the play of colours to advantage.

Amazon Stone.

The stone bearing this name is also a variety of felspar, but its colour is bluish-green. It is slightly translucent at the edges. It is chiefly obtained from Lake Baikal, in Siberia, generally in small pieces, but sometimes large enough to be made into small vases and other ornaments. It is almost an Aventurine, resembling that variety of quartz in being composed of silvery spangles in a green base. It possesses a considerable degree of lustre when polished. A larger green variety is sometimes met with.

Moonstone.

This is a transparent or translucent variety of potash felspar (Adularia), containing bluish-white spots, which, when held to the light, present a pearly or silvery play of colour, not unlike that of moonlight. It is chiefly obtained from Ceylon, and is almost too soft to be worked with advantage by the lapidaries. When properly cut and set it has, however, considerable value, forming a very pleasant and good contrast to rubies and emeralds, especially the latter. This moonstone is probably the stone referred to in the following lines from " Lapidarium" :—

> Nor must we pass the *Selenites* by,
> Whose hues with grass or verdant jasper vie ;
> With the lov'd moon it sympathetic shines,
> Grows with her increase—with her wane declines ;
> And since it thus for heav'nly changes cares,
> The fitting name of "sacred stone" it bears.
> A powerful philtre to ensnare the heart,
> It saves the fair from dire consumption's dart.
> Long as the moon her wasted oil repairs,
> To pining mortals these effects it bears ;
> Yet ne'ertheless, when Luna's on the wane,
> Men from its use will divers blessings gain.
> This stone, a remedy for human ills,
> Springs, as they tell, from famous Persia's hills.

The *Sunstone* is another variety of felspar, transparent when looked at in one direction, but showing by reflected light small golden spangles, owing to the presence of minute scales or crystals of a mineral called *Gœthite*, which is a hydrated[1] peroxide of iron. It is not often used as an ornament.

Serpentine.

This stone is a hydrated silicate of magnesia, and its name is derived from its fancied resemblance to the markings on the skin of a serpent. It is occasionally crystalline, but in that case takes the form of crystals of other minerals, occupying their place when these have been removed by some chemical change. More usually it is massive and opaque, the edges being slightly translucent. It is soft, but tough, easily cut, light, slightly greasy to the touch, and its colour green of various shades, passing into yellowish-brown or yellow. It occupies veins in crystalline rocks, and is almost always in considerable quantity in the few places where it occurs. Among these are the Lizard rock, in Cornwall, and Connemara, in Ireland. It is also common at Zöblitz in Upper Saxony and in Siberia.

Precious or noble serpentine is the name given to the purer varieties. They take a high polish, and are often streaked with white, but it is difficult to obtain pieces of large size free from flaws, which, though not at first defacing the stone, soon undergo change and spoil it. Common serpentine is used for vases, tables, columns, slabs, and chimney-pieces, and occasionally for smaller ornaments. It is very seldom

---

[1] Minerals containing water as an essential ingredient are said to be *hydrated*, from the Greek word ὕδωρ, water.

indeed that the larger slabs are perfect, but such objects as fonts and other church decorations have been made of it from time to time.

Besides this rich and beautiful stone there are many varieties less adapted for ornamental purposes, and known to mineralogists under different names. The purer kinds, those used for personal ornaments, are generally of a rich oil-green colour.

Ophiolite is a name given to include the group of serpentine rocks, and common serpentine is sometimes called *Ophite*. Both are derived from the Greek word for serpent (ὄφις).

Nephrite or Jade.

This curious material is also essentially a silicate of magnesia and lime. It is found in compact masses of leek-green or olive-green colour, passing into grey and greenish-white. It is translucent and remarkably tough, breaking with a peculiarly coarse splinter. It is not very heavy, feels somewhat soapy, like other magnesian minerals, and were it not for the difficulty of working it would hardly have been accounted valuable. It is, however, a very favourite stone in the East, especially among the Chinese, who manufacture it into an infinite variety of grotesque objects and many that are useful.

The names are derived from Greek and Arabic or Spanish words signifying *kidney*, owing to the common use of the stone in the middle ages as an amulet worn to prevent diseases of that organ.

The wide distribution of stones of this kind made

into hatchets, axes, and utensils or ornaments, and found buried beneath the surface, not only throughout Europe and Asia, but even in the islands of the Pacific, including New Zealand, would seem to indicate not only a wide spread of the stone, but an early use of it owing to its toughness. It was not, however, employed for sculpture by the Greeks or Romans, probably for the same reason, as it is not very well-calculated for those delicacies of detail for which classical artists are more remarkable than for the mere overcoming of a mechanical difficulty, which, however, they could perfectly well conquer when the material was of intrinsic value.

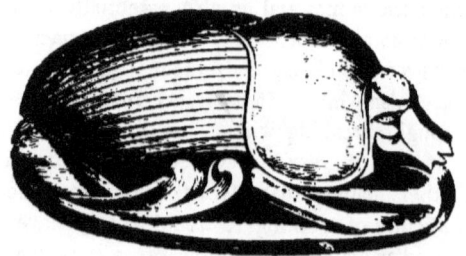

SCARABEUS SCULPTURED IN JADE.—AN EGYPTIAN AMULET.

The only known source of Jade is in the mountain range North of Cashmere, where it has probably been quarried on the same site for some thousands of years. It appears to have been carried to distant countries by the migratory tribes of Central Asia at a period long antecedent to the occupation of the Western countries of Europe by civilized man.

Steatite.

This is another silicate of magnesia, but without lime, and is only a massive variety of talc. It differs in the most remarkable way from nephrite, the one being hard and exceedingly tough, and the other (steatite) one of the softest of stones, yielding readily to the nail and capable of being readily cut into any shape with a common knife. Like jade, but to a greater extent, it has a greasy or soapy feel, and is often called *Soapstone*. Notwithstanding the facility of working it has a close texture, and will take a polish. It is found of various tints of white, grey, yellow, green, and red, the latter colours being due to the presence of a little iron. It is found in many parts of all the principal British islands, but is most abundant in New England and other adjacent parts of North America. It is also very common in the East.

In China there is a stone of the same composition, except that it contains potash to the extent of more than six per cent. This stone is called *Agalmatolite* (from a Greek word signifying image). It is much used in the manufacture of grotesque figures and chimney ornaments which are sold at Shanghai at a very small price. Steatite is of no value for jewelry, and as an ornament is seldom seen, but is extensively used on a large scale for lining furnaces and stoves, as it resists a very intense heat without change. It is said that the Arabs really employ it instead of soap to soften the skin in the bath, and Humboldt describes it as eaten in the absence of food by a tribe of Indians dwelling on the banks of the Orinoco. Taken into the stomach it no doubt prevents the sensation of extreme hunger, but can of course have no value as food. There is considerable resemblance between the serpentines and the steatites, but the latter are rarely used for ornamental purposes and have comparatively little value.

Vast numbers of Scarabæi or Beetles, sculptured of steatite, have been found in tombs and wherever ancient monuments have been disturbed in Egypt. They are generally well cut, and besides beetles there are sometimes more important figures.

Iceland Spar.

This name is given to perfectly transparent and well-crystallized specimens of Calcite or calc spar of which the best are obtained from Iceland.

Calcite or Calcic carbonate is universally distributed throughout the world, and in its various forms is known by several names. It occurs crystalline in about 800 described varieties of form, of which Iceland spar is the most complete, but all are readily reducible to the rhomb. *Iceland Spar* is transparent and colourless when pure, but is often tinted owing to the presence of foreign matter. Good specimens of the spar are absolutely clear, colourless, and transparent, and they exhibit double refraction in a high degree, a black point or line seen through the crystal in a certain direction appearing to be multiplied into two, occupying different places. When heated on charcoal the crystal shines with intense brightness until the carbonic acid is expelled, when it becomes converted into lime. It is too soft to be used for ornaments for personal wear, and is chiefly valuable for optical purposes. Some varieties shine with a phosphorescent light when laid on a heated plate or struck in the dark. Besides Iceland there are many localities from which this mineral is obtained, but nowhere are fine crystals so abundant.

Marble.

VIEW OF ATHENS.—HYMETTUS AND PENTELICON IN THE DISTANCE.

The crystalline limestones not transparent and not showing planes or angles indicating crystalline structure may be included under this head, but we must limit our notice to those that are chiefly made use of for ornamental purposes. The variety even thus limited is, however, very great.

Of all marbles those most valued for statuary purposes are the Parian and the Carrara, both well-fitted for the chisel, and each having yielded some of the most perfect works executed by the human hand. The Parian employed by the ancient Greeks was obtained, as its name suggests, from quarries in the island of Paros, in the Greek Archipelago. These quarries are not now open, and it is difficult to say from their present state whether proper blocks might be obtained for the sculptor. This marble, however, though unrivalled in the soft waxy surface presented by it, and the high and perfect polish it can take, is apt to assume a yellowish tint, and has considerable hardness. The blocks from the celebrated quarries of Carrara, near Spezzia, in Italy, are obtained more readily—the facilities of selection are much greater, and the colour is more purely white than the Parian and less liable to change. The material also works much easier and more pleasant under the chisel, and the polish is very good, if not quite equal to that of the other kind. The difference between the two varieties is one of crystalline texture; the Parian consisting of small facets formed of little crystalline plates applied to one another in every position, while the Carrara is granular, resembling loaf-sugar. The former is called *foliated*, and the latter *saccharoid*.

The Pentelic marble, obtained still from a hill near Athens, resembles the Parian, but is denser and finer grained, besides having small greenish strata of a kind of talc. The celebrated Medicean Venus is of Parian, and the well-known Elgin marbles from the Parthenon, now in the British Museum, are of Pentelic marble. The marble from Mount Hymettus, also near Athens, is another variety, more resembling Parian. The island of Lesbos produced in ancient times a highly-valued statuary marble of the same kind as Parian, but of an almost snow-white colour and very fine grain. All these were comparatively hard in working. There are in Italy pure white marbles of the same kind, though somewhat softer, especially one at Luni, on the coast of Tuscany, which was highly valued by ancient sculptors. India yields some foliated marbles of exquisite tone and smoothness, but also rather hard and yellowish.

The saccharoid marbles of which Carrara is the type are not only of a very pure white, but, as we have seen, retain this colour better than the foliated varieties, which are apt to become grey and in time yellow. The quarries at Carrara appear to have been first opened in the time of Julius Cæsar, and were always highly valued. There are, however, frequent grey veins in this marble, which render it difficult to secure large blocks without a flaw. The quarries are large, and the number of hands employed very great, but the difficulty of obtaining and removing large blocks is still considerable.

White marbles with grey veins are very common in Italy, and have been used extensively for local buildings. Good statuary marble has not been found very abundantly in other countries than Italy and Greece, or if found there has been no certainty of obtaining a sufficient quantity to justify the large outlay required for opening a quarry, supplying machinery, and constructing roads. Very beautiful specimens have been exhibited from time to time from

Spain, Portugal, the south of France and Algeria. The Pyrenees are particularly rich, but hitherto no large supply has been opened up for the sculptor. In the British islands beautiful foliated specimens of the purest colour have been found in Ireland, and a good white marble has been shown from Sutherlandshire, but neither of them has entered the market for general supply.

Black marble, taking a perfectly good polish, without dull stains and without flaws, is perhaps even more rare than the fine white varieties, but being in demand only to a limited extent, it has not assumed equal importance in the market. Fine black marbles, perhaps the finest of all, are obtained from Derbyshire, and to a small extent from Yorkshire. Ireland also contains some. Italy yields black marble, but although the *Nero antico*, the old black, was very perfect, it is almost impossible now to find large slabs in that country free from flaws and capable of taking the highest polish. The high value of slabs of pure black marble arises from their use in the manufacture of Florentine and Roman mosaic work.

Of variegated and coloured marbles the varieties are infinite, and they are found in almost every country. Greece, in former times, and somewhat later, Italy, yielded the finest kinds and almost all colours. Of these, the simple colours are the most pure and the best. The *Rosso antico*, *Verde antico*, and some others well known, were, however, porphyritic breccias, and not marbles in the proper sense.

Among the best known and most valuable marbles

are the Siena marble, or *Brocatello*, a rich yellow in large irregular patches, surrounded by bluish-red or purple veins, and *Cipolino*, a marble with greenish zones, due to the presence of serpentine or talc. All the green marbles are more or less of this nature, that of Connemara, in Ireland, and the Lizard being examples. *Giallo antico* is a very beautiful and rich yellow marble, but it is now rare. Many varieties of variegated marbles are found in England, both in Devonshire and Derbyshire, and many kinds, chiefly greys, are obtained from the carboniferous or mountain limestone of the Penine hills in Yorkshire and Lancashire, and in Westmoreland. Of these, coralline limestones, shell limestones, and encrinital limestones are often when polished sold as marbles.

Lumachelle, or Shell-marble, owes its beauty to fragments of certain peculiar kinds of fossil shells, having pearly iridescence, embedded in and forming part of them. The Ear-shell (*Haliotis*), the Ammonite, and some kinds of Top-shells (*Trochus*) are of this kind. Sometimes green, blue, deep red, and orange tints are reflected, in which case the stone is called *Fire-marble*.

Landscape Marble, also called *Ruin Marble*, or *Cotham Marble*, is a grey limestone found at Cotham, near Bristol, slices of which, cut and polished, present fanciful representations of landscape, trees, and ruins. It is essential that the stones should be sawn at right-angles to the bedding. Similar stones are found near Florence.

Mexican and Egyptian Onyx.—One of the most beautiful, and beyond comparison the most valuable of the almost innumerable varieties of limestone and marble, for ornamental sculpture and for house and church decoration, has been formed in caverns or in fissures in limestone, by the constant dripping of water through narrow cracks in the rock. The water absorbing in its slow passage through and amongst the limestone a quantity of calcareous matter (bi-carbonate of lime) often emerges in drops, which for a time remain on the roof without falling. Undergoing evaporation during this time, a film of carbonate of lime is deposited before the water drops, and this succession of films becomes in time a coating of perceptible thickness on the roof, which as time goes on thickens still more, and at last lengthens and becomes a sort of pipe suspended vertically below the roof where the water drips. The water that falls on the floor likewise evaporates and leaves a film, which rises in time into a column. The pipe descending from above and the column rising from below tend to meet, and in time often do meet, making picturesque columns and sometimes sheets of pellucid limestone for which certain caverns are remarkable. In some cases caverns become completely filled with these stones.

The dropping pipe or upper part of a column thus formed is called a stalactite, and the rising column near its base a stalagmite. The limestone of which they are formed being deposited in successive films has a peculiar structure, easily recognised, sometimes resembling the onyx, which, indeed, has been formed in all probability in a similar manner by the evaporation of water containing silica. The brown stain separated by sharp lines, so remarkable in these Mexican onyxes, is due to the presence of iron oxide.

Although stalactites and stalagmites are common enough in all limestone caverns, they are not always formed under favourable conditions, and it is only occasionally that they are sufficiently large, regular,

STALACTITES AND STALAGMITES.

hard, and tough enough to be valuable for ornamental purposes. The large and perfect blocks adapted for cutting are very valuable, and are rarely met with. They are obtained in greatest perfection from the Pyrenees and Egypt, and in some caverns in South America, hence the local names, Egyptian and Mexican Onyx, by which they are recognised.[1] Stones such as these are sometimes called *Oriental Alabaster* and *Alabastritis*, their transparency and colour being not unlike alabaster. They have no relation whatever to true alabaster, which is much softer, and is a sulphate of lime, a totally different mineral from the carbonate, and formed under different conditions.

Some fine specimens of these very beautiful varieties of stalactitic or stalagmitic limestone have been worked up into ornamental forms. There is a noble sarcophagus in the Sir John Soane Museum, constructed by the Egyptians, at a very early period, out of a single block, and covered inside and outside with hieroglyphics. It was purchased for two thousand guineas. Columns of large size, measuring 40 feet in length, of a single stone, have been obtained within the present century, and tables and other articles of furniture were sent from Rome to the Great Exhibitions of 1851 and 1862, showing much ingenuity in adapting the markings of the stone to the purposes of the work. Lately it has been introduced into England for chimney-pieces and house and church decoration. Its peculiar brown colour does not adapt itself to every style of ornamentation, and it requires some ingenuity and taste to do full justice to its beauties.

There can be no doubt that the following account, by Pliny, refers to the stone we are now considering,

[1] There are also some quarries in the province of Oran in Algeria whence the ancients obtained this stone.

THE SOFTER GEMS AND VALUABLE STONES. 149

and it is probable that the perfume jar containing the ointment with which our Saviour was anointed previously to His trial and execution was of the same material: "This Onyx stone, or Onychitis aforesaid, some name Alabastrites; whereof they use for to make hollow boxes and pots to receive sweet perfumes and ointments, because it is thought that they will keepe and preserve them excellently well without corruption. The same being burnt and calcined is very good for diverse plastres" (Bk. xxxvii. chap. 8).

"Alabastrites" was a material much used by the Romans, for the purpose of containing *unguenta* or perfume oils. In these cases the neck of the vessel was broken off when its contents were required, as it had been hermetically sealed by the maker to prevent the evaporation of the scent. In the museum at Naples are some large jars of this kind, made of the same material, still retaining a strong perfume from their former contents. They were found at Pompeii. There are also in the Naples and other museums sacred Egyptian vases, with a cover shaped like the head of a mummy, and constructed of the same material.

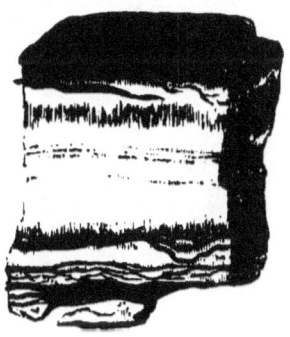

SATIN SPAR.

Satin Spar, of which the annexed cut gives a

sufficient idea, is a white and fibrous variety of limestone, moderately hard, and capable of being cut into snuff-boxes, turned into studs, and adapted in other ways to ornamental purposes. It is very different in appearance from marble, but it is a true carbonate of lime, and possesses crystallization of a certain kind. The same name is applied to a variety of alabaster which, like Egyptian onyx, it much resembles (see p. 152). **Alabaster.**

SCULPTURED VASES IN ALABASTER.

The stone named *Alabaster*, in modern times, is a

sulphate of lime, obtained in large quantities near Volterra in Tuscany, but also met with in abundance, though rarely in pieces fit for ornamental work, in our own country, in France, and in many other parts of Europe. The stone found near Volterra is of a rich brown colour, passing into lemon-yellow, and exists in large blocks. It is very soft, and cuts into any shape with extreme facility. It is soluble in water, though only to the extent of one part in five hundred. It is translucent, and is sometimes manufactured into lamps, the light piercing through the thin walls of the lamp with a soft and beautiful radiance.

Fine quarries of alabaster existed in Assyria, and

ASSYRIAN OR BABYLONISH CYLINDERS.

were certainly resorted to for ornamental purposes. Among the objects most frequently sculptured were cylinders, on which inscriptions are very common.

A good idea of the nature of these, and of one of the ancient uses of this mineral, will be better understood by the annexed illustration than by any lengthened description.

When in a crystalline form, sulphate of lime is called *Selenite*, from the Greek word σελήνη, *the moon*, perhaps from an idea once entertained that it consisted of water congealed by the moon's influence, or perhaps from its reflection of light, which is peculiar. The crystal splits very easily with a knife into thin plates or sheets, which, being perfectly transparent, can be used as glass. The late Mr. David Forbes describes the windows of a church in Bolivia as being formed of this material in slabs about two inches thick. It is also generally used in that part of South America for window-panes.

The name of *Satin Spar* (see p. 149) is given to fibrous varieties of massive selenite, found in Derbyshire and Nottinghamshire, and sometimes in Gloucestershire. A stone bearing this name is cut and set in rings, and then resembles the cat's-eye, but is very much softer. It is also manufactured into various table ornaments.

The massive form of sulphate of lime is called *Gypsum*, the chief value of which is for burning into plaster-of-Paris, which sets very rapidly when mixed with water and put into a mould. The following interesting account of its use by the ancients is quoted from an English translation of Theophrastus by Sir John Hill:—

"The stone from which gypsum (plaster-of-Paris)

is made by burning is like alabaster; it is not dug, however, in such large masses, but in separate lumps. Its viscidity and heat, when moistened, are very wonderful. They use this in buildings, casing them with it, or putting it on any particular place they would strengthen. They prepare it for use by reducing it to powder, and then pouring water on it, and stirring and mixing the matter well together with wooden instruments; for they cannot do this with the hand because of the heat. They prepare it in this manner immediately before the time of using it, for in a very little while after moistening, it dries and becomes hard, and is not in a condition to be used.

"This cement is very strong and often remains good, even after the walls it is laid on crack and decay and the sand of the stone they are built with moulders away: for it is often seen that even after some part of a wall has separated itself from the rest and is fallen down, other parts of it shall yet hang together and continue firm and in their place by means of the strength of this matter which they are covered with.

"This gypsum may also be taken off from buildings, and by burning again and again be made fit for use. It is used for the casing of outsides of edifices, principally in Cyprus and Phœnicia; but in Italy for whitening over the walls and other kinds of ornaments within houses. Some kinds of it are also used by painters in their business; and by the fullers, about cloths.

"It is also excellent and superior to all things for

making images; for which it is greatly used and especially in Greece because of its pliableness and smoothness."—Theophrastus, cxii. to cxvi.

Gypsum in the form of plaster is still used very extensively, but in our climate it is not adapted for exposure to the open air. It is largely employed in moulding, and, with some modifications, is employed in making casts of sculpture and other works in stone or metal.

Fluor-Spar or Blue John.

This mineral, which is a fluoride of calcium, or combination of the two elements fluorine and calcium, is found abundantly in many limestone districts, and is used in Derbyshire for ornamental purposes. It is moderately hard and of various colours, such as white, grey, green, and various tints of blue, but its most characteristic colour is a rich dark blue, which bears the name of Blue John. It is also sometimes yellow, purple, or red. It is perfectly clear and generally transparent, but brittle and easily scratched. When pounded and placed on a shovel over a fire it is phosphorescent. With greater heat the phosphorescence ceases, but it may be recovered by an electric spark. When fragments are rubbed together in the dark they become luminous. Some specimens are blue when the transparent crystal is looked through, but appear green when looked at or seen by reflected light.

The specimens of deep blue colour found in Derbyshire are cut into tazzas, vases, obelisks, and other ornamental forms, and are much admired. Large fragments generally appear full of cracks internally. The colour is often modified artificially by the application of heat.

This spar is usually found in well-developed crystals, which are generally perfect cubes striated as shown in the subjoined woodcut. Occasionally large blocks of fluor spar are found in which the crystalline form is not indicated.

CRYSTALS OF FLUOR SPAR.

CHAPTER VI.

VALUABLE MINERALS DERIVED FROM THE VEGETABLE
AND ANIMAL KINGDOMS :—JET, AMBER, PEARL,
CORAL.

BESIDES the stones valued for ornamental purposes and strictly belonging to the mineral kingdom, there is a small number consisting of vegetable or animal substances still retaining every mark of their origin, but existing in a form so solid and so permanent as to entitle them to rank as stones. We do not include among them what are called petrifactions or accumulations of fossil shells and other substances sometimes met with—these belong rather to the varieties of limestone or quartz—but there are four somewhat remarkable and well-defined substances that unmistakably belong to the class we are now establishing.

Jet.

The first of these that comes under consideration is that variety of coaly matter found buried in certain rocks among which plants and animals have lived. It is, however, of so different a nature from common coal as to be employed for many purposes as an ornamental stone.

The name jet is a corruption of Gagates, its ancient

appellation, derived from the river Gaga, or the town of Gagis, in Lycia, where the mineral was originally found. In former times it was chiefly used in medicine, and was employed in magic as a means of incantation. It was also used for staining pottery. Anklets and bracelets made from it by the early British inhabitants of our island have, however, been found, proving its early use for ornamental purposes; and there were discovered in 1846, in two stone coffins deposited under the chief entrance of the church of St. Geréon, in Cologne, as many as twenty-six articles manufactured of this material, forming a complete suite of jet ornaments. The ornaments consisted of two hair-pins, with heads composed of pine cones, almonds, and trefoils, bracelets, rings, and a half crotalon with a head of Medusa. They are supposed to have been the ornaments of some priestess of Cybele. In a Roman tumulus near Whitby, in Yorkshire, a jet earring of a lady in the form of a heart was found lying close to the jawbone of a skeleton. In this heart there was a hole in the upper end for suspension in the ear. Whitby is well-known as yielding large supplies of jet, and there is no doubt that when the abbey of Whitby existed as a seat of learning and the abode of pilgrims, jet rosaries and crosses were common. The manufacture of church ornaments was carried on there actively till the time of Elizabeth, but then ceased till the beginning of the present century. Until this latter date jet was only roughly cut with knives and files, but the use of the lathe being then introduced, a much improved manufacture was started which has since become very important. The value of the trade to the town of Whitby now approaches £100,000 per annum.

Jet is of a velvet-black or brownish-black colour, with a brilliant and resinous lustre, breaking with a smooth, conchoidal, or shell-like fracture, and often dividing into regular prismatic shapes. It is smooth, and does not stain the fingers. It is light, but heavier than wood, sinking in water. It burns with a greenish flame, emitting a strong, sweetish, bituminous smell, and leaving a pale ash. It is opaque, but in thin slices generally shows a woody texture.

The precise history of this mineral is still doubtful, though it is always derived from either animal or vegetable life, and may often be identified with vegetable structure. It is found not only in lumps and beds, but outside flint pebbles, encrusting them, seeming to show that it had once been liquid. Besides flint pebbles, both wood and bone have been found encrusted by it.

In his "Geology of the Yorkshire Coast," Dr. Young, who was a very careful observer, writes as follows: "Jet, which occurs here in considerable quantities in the aluminous bed, may be properly classed with fossil-wood, as it appears to be wood in a high state of bitumenization. Pieces of wood impregnated with silex are often found completely crusted with a coat of jet about an inch thick. But the most common form in which the jet occurs is in compact masses from half-an-inch to 2 inches thick, from 3 to 18 inches broad, and of 10 to 12 inches long. The outer surface is always marked with longitudinal striæ, like the grain of wood, and the transverse fracture, which is conchoidal and has a resinous lustre, displays the annual growth in compressed elliptical zones." It must not, however, be concluded that all jet is of woody origin.

Although jet is common in the alum shales of the Yorkshire lias, being found there in great abundance, it is not found in those parts of the country where large accumulations of vegetable matter have caused the formation of coal, and there are many parts of the world where it is also met with. In France, both in the Ardennes in the north and the Pyrenees in the south, large quantities are worked, and it is said that in the last century not less than 1,200 men were employed in the department of Aude alone in carving and turning the jet of that neighbourhood into beads, rosaries, buttons, bracelets, earrings, necklaces, snuff-boxes, drinking-vessels, and pieces cut into facets for mourning ornaments. Fifty tons weight were actually consumed in this way in a year, but the trade has since greatly fallen off. There has also been a large quantity of jet found in Silesia and on the shores of the Baltic. Spain yields considerable supplies.

Near Whitby, the shales, or slaty rocks, are carefully examined for very thin laminations of jet, which, being followed, lead to thicker and more available deposits. The best and most productive shales are in the lower part of a series now called the "jet-rock series," overlying the alum shale. Here there is one

bed about 20 feet thick, known as the "jet-rock.'
It is chiefly developed in the estates of Lord Mulgrave,
but the rock extends both north and south of Whitby
for some miles. There is a reference to this locality
in Drayton's curious poem of " Polyolbion :"

> The rocks by Moultgrave, too, my glories forth to set,
> Out of their crannied nooks can give you perfect jet.

There are two kinds of jet, the hard and the soft,
but the former is the most valuable, and much of
the jet now worked up at Whitby comes from Spain.

Besides its use for small jewelry, jet was once
celebrated for its medicinal and other virtues. The
following lines from Marbodus, whose poem has
been so often quoted, will show the extent of superstitious belief in the Middle Ages with regard to this
material :—

> Lycia her jet in medicine commends ;
> But chiefest that which distant Britain sends :
> Black, light and polished, to itself it draws
> If warmed by friction, near adjacent straws.
> Though quenched by oil, its smouldering embers raise
> Sprinkled with water a still fiercer blaze :
> It cures the dropsy, shaky teeth are fix'd
> Wash'd with the powder'd stone in water mix'd.
> From its deep hole it lures the viper fell,
> And chases far away the powers of hell ;
> It heals the swelling plagues that gnaw the heart,
> And baffles spells and magic's noxious art.

A great deal of artificial jet is sold for ornaments,
and in some cases is passed off for the real. It is
however, much heavier and less brittle. It is a kind
of black glass cut into facets, or blown into beads,

and black wax is used either to fasten the glass to an iron back or to fill the beads.

Some kinds of *cannel coal* are occasionally worked into ornaments and sold as jet. These minerals, however, have far inferior lustre to the real jet, and are much less adapted to serve as ornaments. Varieties of *lignite* are made use of for similar purposes. Veins of a peculiar kind of jet are not unusual in beds of lignite, but they are of small thickness and cannot be worked into shape.

Jet is a combination of carbon and hydrogen, and differs little from coal in its essential properties. Of the varieties of coal it most resembles anthracite. It is regularly mined, and to such an extent that the waste thrown out on the hill-sides disfigures the neighbourhood when the works are carried on inland. The source would seem to be rather bitumen than any woody matter; wood, scales of fishes, or other foreign bodies traceable in it having been converted into jet. Whether the bitumen was obtained by the decomposition of animal or vegetable matter is uncertain and unimportant. The mining operations are, like other mining, very speculative, the jet not lying regularly, and the profits of one season being often wasted in the unsuccessful operations of the next. The largest seam discovered of late years weighed 370 stone, and was worth £250.

Amber.

Amber, which, if not to be regarded as a stone, must still be accepted as one of the precious gems referred to the mineral kingdom, is generally admitted to be the mineralised resin of some extinct coniferous trees resembling pines that once grew on the plains of northern Europe in great abundance and extended through Europe even so far as the shores of the Mediterranean Sea. It is chiefly found on the shores of the Baltic between Königsberg and Memel, but has been met with in Greenland, France, England, Switzerland, and Sicily. It is found in irregular masses of yellow colour, varying in different specimens from the palest primrose to the deepest orange or brown. It is brittle and harder than gypsum, though scratched with calcite. It yields readily to the knife, and breaks with a smooth fracture. It burns with a yellow flame, emitting a pleasant odour, and leaves when burnt a black residue, coaly and very light. It can be dissolved in spirits of wine, and is very slightly heavier than water.

Of all substances amber is that in which electricity is most easily excited by friction, and this has long been known and recognised, as its name among the ancients was *electron*, whence the word electricity is derived.

Large quantities of amber are cast ashore in Pomerania and adjacent parts of the Baltic coast, during the storms that disturb the waters of that inland sea at the time of the equinox, and the search for these specimens though precarious is an important industry exercised by the dwellers on the shore between Dantzic and Memel. All along the coast there are deposits of alluvial sand and clay, mixed up with

MINERALS OF ORGANIC ORIGIN. 163

fragments of fossil-wood and other minerals, and a considerable quantity of amber is also obtained by mining in these beds at favourable seasons. From time to time considerable deposits are come upon. "The amber-fishers, clothed in leather dresses, wade into the sea and seek to discover the amber floating on its surface, which they secure with bag nets hung at the ends of long poles. They conclude that much amber has been detached from its bed when they discover many pieces of lignite floating about. Mining is carried on by sinking through the sand and superficial strata to the beds containing the amber and lignite; many of these pits are sunk to the depth of 130 feet. The faces of the precipitous cliffs are explored in boats, and masses of loose earth or rock supposed to contain the object of search are detached with long poles having iron hooks at their ends."

Specimens are obtained from time to time all along the English shores of the German Ocean, especially at and near Aldeburgh. Others have been found in the river gravels, and in older gravels near London.

The lumps of amber generally found are of small size, often not larger than a hazel-nut, but occasionally very large pieces are obtained. A mass weighing 13 lb. was on one occasion picked up, and its value was estimated at not less than £4,500. A good piece weighing one pound is worth nearly £8.

Among the ancients amber was said to consist of the tears of those trees to which the sisters of Phæton were transformed, after their brother had tried in vain to conduct the car of Apollo. It was hence called *Succinum*, from *succum*, the juice of a tree. It is also sometimes supposed to be referred to as the *Lyncurium*, and was said to be formed of the excreta of the lynx; but the stone more usually bearing that name was the Hyacinth, which has been already described, and to it the description more clearly points. The modern name is probably derived from the Arabic or Persian language, and resembles that by which it is known in the East.

With more reason, perhaps, than in many cases, the ancients attributed to the amber great virtue in medicine, and other singular properties. It was believed to be a preservative against all complaints of the throat, and for this reason was worn by women and children. Pliny states that a collar of amber worn round the neck of a young infant was looked on as able to prevent secret poisoning, and as a countercharm to witchcraft and sorcery:—"Callistratus saith, that such collars are good for all ages, and namely to preserve as many as weare them against fantasticall illusions and frights that drive folke out of their wits: yea, and amber, whether it be taken in drink or hung about one, cureth the difficulty of voiding urine."—(Holland's Pliny, Bk. xxxvii. ch. 3.)

That amber is of vegetable origin is very certain, both by its chemical composition and its optical properties; and that it was once in a liquid state is clear from the condition in which insects and the remains of insects, along with fragments of leaves and stalks, are found occasionally buried in it.

Certain insects, such as have four membranous wings, as the bee and wasp, are rare; and insects with two wings, as gnats and flies, are more numerous. Then come the spider tribe, and some beetles, chiefly those that live on trees. In many cases they appear to have been caught in a viscous fluid, and struggled to extricate themselves, occasionally leaving behind them, at a distance from the body, a leg or a wing. It has been stated by entomologists that the insects enveloped in amber are in general such as haunt the trunks of trees, or live in the fissures of their bark. They are not of the same species as the insects now living in northern Europe, and resemble species inhabiting warm rather than temperate latitudes. Mr. Hope, who has described them minutely, and is the best authority on the subject, states them to be altogether extra-European. Besides insects and leaves, other objects have been occasionally found buried, among them being, in one case, the leg of a toad.

Amber is largely used in the East for the construction of mouthpieces to pipes, and forms an indispensable part both of the meerschaum and the tchibouk. In Turkey there is a current belief that amber is incapable of transmitting infection, and as in that country it is a mark of politeness to offer the pipe to a stranger this important negative peculiarity is the more appreciated. Not only the Turks, but the Egyptians, Arabs, Persians, and natives of India, all ornament their pipes in this way, and some idea of the extravagance of the taste may be obtained when it is known that four amber mouthpieces, set with brilliants, exhibited in the Turkish section of the Great Exhibition of 1851, were valued together at £1,000.

But it is not only mouthpieces of pipes that absorb large quantities of the amber sent to Constantinople and India from northern Germany; the arms, the saddles and bridles of horses and camels, and even the tubes of pipes, are sometimes covered with specimens of it. In various parts of Europe amber is made into brooches, ear-rings, boxes, cane-handles, and even occasionally into salvers, candlesticks, and other large articles. One very important use is its manufacture into beads of all kinds. Amber necklaces are greatly admired, and looked on as among the principal ornaments of large classes. The Russian peasant-girls are often seen wearing two, or even three, rows of beads, presumed to be of amber, though, perhaps, not unfrequently imitations.

Amber beads are often cut in facets, and occasionally are large and very beautiful, but though long in fashion in Western Europe, they are now rarely worn pierced and as necklaces, except among peasants. Another use of amber, now almost out of fashion, was as cane-tops, especially used by the fops of the early part of the last century. The canes employed were slender and fragile, and used entirely for ornament. They were often richly mounted with silver, gold, amber, or precious stones. In the Mineralogical Museum in Paris there is the handle of a cane made of amber, the colour of which is so pure a yellow and so limpid that it might almost be mistaken for a Brazilian topaz. Pieces of amber of the large size and perfect clearness of this specimen are rarely met with.

The clouded varieties are for some purposes more highly valued even than these perfectly clear and brilliant specimens. The straw-yellow, slightly clouded, translucent variety, when of good size, is purchased by the Orientals at extravagant prices.

The working of amber into fanciful shapes is difficult, owing to the extreme electrical sensibility of the material, and during the process of polishing it frequently becomes so excited as to crack and fly to pieces. It is usual to cut the amber on the lathe with a leaden disc, and polish it on the lathe with whiting and water, or with Trent sand or scraped Flanders brick and water, or with rottenstone and oil. When the specimens have been turned into shape at the lathe they are polished with glass-paper and rottenstone; but in polishing it is usual for the workman to have several specimens in hand at once, taking first one and then another in turn, leaving them at rest during the interval to avoid continued friction and the consequent accumulation of electricity. It is even said that some of the men are troubled with nervous tremors in their wrists and arms, from the electricity developed during the operation of amber-working.

Amber may be joined together when broken, or two pieces may be united firmly and permanently by smearing the clean edges with linseed oil and pressing the parts together, while held over a charcoal fire.

A cement is sometimes used for joining broken fragments, composed of linseed oil, gum-mastic, and litharge. Some workers in amber simply warm the fractured surfaces and press them together, after moistening them with a solution of potash or soluble glass, the pieces being tied round with a string for a few days.

Amber is imitated by mixing gradually, at a moderate and gradually raised heat, rectified oil of asphalte with turpentine, in a copper or brass vessel. After two or three boilings it becomes thick, and may be poured into moulds.

Besides being used as an ornament, amber is sometimes distilled to obtain succinic acid (which crystallizes in needles on the dome of the still) and the oil of amber. The latter drops from the beak of the still into a receiver. The operation is carried on at a low temperature, and the residue is an extremely black shining substance, used as the basis of the finest black varnish.

Amber is a combination of about 80 per cent. of carbon, 10 per cent. of hydrogen, and 10 per cent. of oxygen, which is nearly the same as that of the vegetable resins.

Pearl.

Pearl is the name given to an unnatural growth of the nature of a spherical or oval protuberance on the inner side of the shells of certain molluscous animals living in the water. These protuberances have a peculiar transparency and play of colours called

MINERALS OF ORGANIC ORIGIN. 169

iridescence, from its presenting the various colours of the rainbow (*iris*). Many shells exhibit this iridescence, which is called *Mother-of-pearl*, or *nacre*, and true pearls occur only on the shells which have this property. Being very rare, and often extremely beautiful, pearls have always been very highly valued for ornamental purposes.

Formed entirely by the animal during its growth, and the deposit governed by accidents of growth, the pearl is only a mineral in a limited sense of the word; but being altogether independent of the animal when once formed, it may be treated of here for the same reason and in the same sense in which *tabasheer* has been referred to amongst opals, and amber amongst the forms of carbon.

The animals whose shells are nacreous, or consist of mother-of-pearl, are certain kinds of oyster, the Unio, or river-mussel, the Ear-shell (*Haliotis*), some of the Top-shells (*Turbo*), and some other kinds of shells, of which the living nautilus and the extinct ammonite are examples.

The kinds of oyster that contain the pearl inhabit the sea, living on banks at a moderate depth, and are found chiefly in the Indian seas. The river-mussel yielding pearls is common in Scotland and in other places. The other shells named rarely contain valuable pearl. Almost all the pearl-yielding shells are covered when living with a thick skin.

All these shells have a similar structure, being made up of thin plates, separated by, or rather enclosed in a peculiar glue, secreted by the animal, and quite fibrous, so that when the carbonate of lime (the solid part) is removed by an acid this skeleton remains.

The wavy appearance and beautiful play of colours, or, in other words, the nacreous appearance of the mother-of-pearl, is caused by this peculiar structure and the semi-transparency of the animal matter between the plates. The pearls themselves seem to have been thrown out at some place where the shell while in process of deposition has been slightly injured, and the round concretion is formed by the deposit of alternate layers of carbonate of lime and this animal matter.

> The sea-born shell conceals the *Union* round,
> Called by this name as always single found.
> One in one shell, for ne'er a larger race,
> Within their pearly walls the valves embrace.
> Prized as an ornament its whiteness gleams,
> And well the robe, and well the gold beseems.
> At certain seasons do the oysters lie
> With valves wide gaping towards the teeming sky,
> And seize the falling dews, and pregnant breed
> The shining globules of th'ethereal seed.
> Brighter the offspring of the morning dew,
> The evening yields a duskier birth to view;
> The younger shells produce a whiter race,
> We greater age in darker colours trace.
> The more of dew the gaping shell receives,
> Larger the pearl its fruitful womb conceives;
> However favouring airs its growth may raise,
> Its utmost bulk ne'er half-an-ounce outweighs.
> If thunders rattle through the vaulted sky
> The closing shells in sudden panic fly;

Killed by the shock the embryo pearls they breed,
Shapeless abortions in their place succeed.
These spoils of Neptune th' Indian Ocean boasts;
But equal those from ancient Albion's coasts.

The above account of the pearl, though imaginative, is not untruthful. It points out both the principal sources of the gem, and something of the natural history peculiarities for which it is remarkable. It is, however, chiefly true as far as regards the Pearl-Oyster of the Indian seas (a species of *Meleagrina*), not applying so well to the Unios of our rivers. It should be remembered that the presence of occasional thicknesses is common in many shells, although most remarkable in those oysters and fresh-water mussels in which the pearl is most common. Besides Ceylon, the Coromandel coast, and the Persian Gulf, which are the best known sources of supply of pearl-oysters in the East at present, and the coast of Arabia, which was once very celebrated, several places on the coast of Colombia, and others in the Bay of Panama, have yielded specimens in considerable number, though not of late years of the highest purity and size. It is said that Seville alone imported nearly seven hundred pounds weight of pearls from America in the year 1587, and a very celebrated gem which belonged to Philip II. of Spain, and weighed 250 carats (more than two ounces), was brought from Margueritʌ on the coast of Colombia. This pearl was valued at 150,000 dollars (£30,000). The banks whence these were obtained have been fished over within the last half-century, but with no satisfactory result. The same may be said of the coast of Arabia, where once valuable pearls were got. No doubt here, as on our own shores, the oyster by continual disturbance during the breeding season was at length destroyed, or migrated to some more quiet home.

The oysters that are richest in mother-of-pearl are those likely to yield pearls. They inhabit beds or banks covered with from seven to nine fathoms of water, and nearly out of sight of land. Those near Ceylon are exposed during a great part of the year to strong and often irregular currents, which sweep the bottom and from which no protection can be given, and the weather is generally so bad that a regular inspection of the banks of a fishery can only be carried on during the month of March. When the banks are well-covered with oysters they are sometimes exposed to the attack of skates, which often do a vast amount of mischief and destroy a whole year's produce, the value of which may be guessed at when it is known that in the year 1864 the Ceylon Government anticipated a return of £50,000 from the produce; and realized nothing. Thus it is evident that other causes besides over-fishing may have interfered with the production of pearl-oysters where they were once abundant.

Pearl-oysters attain their full superficial growth in about four years, but the shell thickens for two years more, and during this period the pearls increase rapidly in size. After six years the animal dies, the shell opens, and the contents disappear. It is most desirable to leave the oysters as long as possible, in order to obtain large pearls, but there is the danger of leaving them too long, and losing them altogether.

Tavernier observes: "The more it rains in the course of the year the more productive it is for the fishery; but many persons imagine that the deeper water the oyster is found in so much is the pearl the whiter, because the water is not so hot, the sun not penetrating to the bottom, but this is an opinion I beg leave to contradict. They fish in from four to twelve fathom water, which fishery is carried on upon the banks, where there are sometimes as many as two hundred and fifty barks, in the greater part of which there is but one diver, some of the largest having only two."

The following account of the fishery of the pearl in Ceylon is taken from a work on that island by Captain Percival :—"There is, perhaps, no spectacle in Ceylon more striking than the bay of Condatchy during the season of the pearl-fishery. This desert and barren spot is at that time converted into a scene which exceeds in novelty and variety almost anything I ever witnessed; several thousands of people of different colours, countries, casts, and occupations continually passing and repasssing in a busy crowd; the vast number of small tents and huts erected on the shore with the bazaar or market-place before each; the multitude of boats returning in the afternoon from the pearl banks, some of them laden with riches; the anxious, expecting countenances of the boat-owners while the boats are approaching the shore, and the eagerness and avidity with which they run to them when arrived in hopes of a rich cargo; the vast numbers of jewellers, brokers, merchants, all occupied in some way or other with the pearls, some separating and assorting them, others weighing and ascertaining their number and value, while others are hawking them about, or drilling and boring them for future use,—all these circumstances tend to impress the mind with the value and importance of that object which can of itself create this scene.

"As soon as the oysters are taken out of the boats they are carried by the different people to whom they belong, and placed in holes or pits dug in the ground to the depth of about two feet, or in small square places, cleared and fenced round for the purpose, each person having his own separate division. Mats are spread below them to prevent the oysters from touching the earth, and here they are left to die and rot. As soon as they have passed through a state of putrefaction and have become dry, they are easily opened without any danger of injuring the pearls, which might not be the case if they were opened fresh, as at that time to do so requires great force. On the shell being opened the oyster is minutely examined for pearls, and it is usual even to boil the oyster, as the pearl, though commonly found in the shell, is not unfrequently contained in the body of the fish itself.

"The stench occasioned by the oysters being left to putrefy is intolerable, and remains long after the fishing is over, but this does not prevent the search for pearls accidentally lost being continued long after the end of the season.

"In preparing the pearls, an obtuse inverted wooden cone, about six inches long and four inches across, is used, supported on three feet about twelve inches long. In the upper flat surface holes are made to receive the larger pearls, and the smaller ones are beaten in with a little wooden mallet. The drilling instruments are spindles of various sizes, made to turn in a wooden head by means of a bow-handle. The pearls being placed in the pits, as already described, and the point of the spindle adjusted to them, the workman presses on the wooden head of the machine with his left hand, while his right is employed in turning round the bow-handle. During the process of drilling, he occasionally moistens the pearl by dipping the little finger of his right hand in a cocoa-nut filled with water, which is placed by him for that purpose; this he does with a dexterity and quickness which scarcely impede the operation, and can only be acquired by much practice. They have also a variety of other instruments, both for cutting and drilling the pearls; and to clean, round, and polish them to the state in which we see them a powder made of the pearls themselves is employed."

Pearls of considerable beauty have been found in Scotland, the principal rivers being the Forth, the Dee, the Don, the Tay, and the Tweed. The pearls are brought by the countrypeople to the towns, where they are sold at prices varying from a few shillings up to £25. Scottish pearls are, however, easily distinguished from the fine Oriental pearls, being of a different shade of colour. Their beauty of lustre and form, and their fine opaque colour, are said by Mr. A. Cockburn to attract more attention now than formerly. The late Prince Albert ordered a necklace to be made of pearls of this kind of a certain size, but it took more than twenty years to complete it, so that the stones are not abundant. A fine specimen of Scottish pearl, sent to the Dublin Exhibition some years ago, was valued at £500. It was set in enamel and gold, and ornamented a tiara for a lady's head-dress.

Tavernier describes a pearl belonging to the Imam of Muscat, which though it weighed only $12\frac{1}{16}$ carats, and was not perfectly round, was, in his time, considered to be the finest in the world, because it was so bright and transparent that you could almost see the light through it. The same author mentions a pearl of fifty-five carats which he had in his possession, the shape of which was that of a pear, and the water very clear. This was from the island of Marguerita, near Cubagua, in Colombia. The American pearls were known and highly valued at the time when Tavernier wrote; and he also mentions pearls from Bavaria, and from Scotland, but he considers that although valuable for necklaces, they will not bear a comparison with those from the East and West Indies. The following quaint remarks on the subject of pearls will be read with interest:—

"Before closing this chapter, I wish to make an important remark respecting pearls, and the difference of their water, some being very white, bordering upon yellow, and some again of a blackish or lead colour. With respect to the latter, they are found only in

America, and their colour is owing to the nature of the bottom, which contains more mud than in the East. In a cargo that the late M. du Jardin, the famous jeweller, had in the Spanish galleons, he found six pearls perfectly round, but black as jet, which, taking the one with the other, weighed twelve carats. He gave them to me, in company with other articles, to carry to the East and endeavour to sell, but I brought them back to him, not having been able to find any person who was pleased with them. As to those which have a yellow cast, this arises from the pearl-fishers selling the oysters to the merchants by heaps, who sometimes keep them as long as fourteen or fifteen days until they open of themselves, when they take out the pearls; during which time some of these oysters, losing their moisture, spoil and waste, by means of which infection the pearl turns yellow, which is so true, that in all the oysters that have preserved their moisture, the pearls are always white. The reason why they keep them till they open of themselves is, that if open by force, as we do our shell-oysters, they would run the risk of damaging or breaking the pearl. The oysters at Manar (in Ceylon) open naturally five or six days sooner than those in the Persian Gulf, because the heat is greater at Manar, which is in the tenth degree of north latitude, than at the island of Bahren, which is about twenty-seven degrees; thus among those pearls which come from Manar there are few yellow. In short, all the Eastern nations are exactly of our taste with regard to whiteness, and I have always observed that they like the whitest pearls, the whitest diamonds, the whitest bread, and the whitest women."

The manufacture of artificial pearls is carried on extensively in many countries, and the results are sometimes very beautiful. The French excel in imitative gems of all kinds, pearls among the rest, and have introduced a kind of opaline glass of a nacreous or pearly colour, very heavy and fusible, which gives to beads the different weights and varied forms found among real pearls. Gum is used to fill them, by which great transparency is obtained, and the glassy appearance of the surface is taken away by the vapour of hydrofluoric acid. The apparatus employed in the manufacture is very ingenious. The beads are made out of small glass tubing, drawn to enormous lengths, and then cut into pieces about a foot long. These are afterwards subdivided into short cylinders of equal length and diameter by drawing a knife over a large number of them, laid horizontally in a row on a sharp edge. The cylinders thus obtained are brought into the pearl shape by exposing them to a suitable heat, sufficient to soften them, stirring them all the time. When well rounded, they are separated from the powder with which they are mixed, by careful agitation in sieves, and are polished and finally cleaned by shaking them in canvas bags. The result is extremely beautiful.

Roman pearls are also very good imitations of the real. The method of preparing them is borrowed from the Chinese, and they are coated with a nacreous liquid prepared by throwing into an ammoniacal preparation the scales, or rather the *lamellæ* of the scales, of a small river fish. When digested in ammonia, these scales become soft and flexible, and are held in suspension by the liquor, which however soon evaporates on exposure to the air, leaving the scales behind. A little isinglass is sometimes used to render the scales adherent. The imitation is good and durable, but these imitative pearls are heavy.

Inferior imitations of pearls are made in Saxony of glass globes, coated with wax, but these are cheap, and soon injured on exposure.

Coral.

The beautiful specimens of Coral, fished up in some temperate and warm seas and manufactured into various ornaments in Italy, can hardly be regarded as gems, but they are so highly valued and are so unmistakably stony in their composition that we may well include them among the list of minerals of organic origin. It would not be easy to explain the ancient idea of the nature of coral in a manner more clear and picturesque than in the following extract from a poem by Orpheus, already quoted and referred to in Mr. King's "Ancient Gems," p. 423:—

> The Coral, too, in Perseus' story fam'd,
> Against the scorpion is for virtue nam'd;
> Above all gems in potency 'tis rais'd
> By bright-hair'd Phœbus, and its virtues prais'd:
> For in its growth it shows a wond'rous change—
> True is the story, though thou'lt deem it strange.
> A plant at first, it springs not from the ground,
> The nurse of plants, but in the deeps profound.

Like a green shrub it lifts its flowery head
Midst weeds and mosses of old Ocean's bed.
But when old age its withering stem invades,
Nipped by the brine its verdant foliage fades;
It floats amid the depths of ocean toss'd,
Till roaring waves expel it on the coast.
Then in the moment that it breathes the air
They say, who've seen it, that it hardens there.
For as by frost congeal'd and solid grown,
The plant is stiffen'd into perfect stone;
And in a moment in the finder's hands
Late a soft branch, a flinty coral stands.
Yet still the shrub its pristine shape retains,
Still spreads its branches, still the fruit remains.
A sweet delight to every gazer's eye,
My heart its aspect fills with speechless joy.

But this account hardly agrees with that of the naturalist of modern times, whose closer observation is not so liable to be warped by the exercise of the imaginative powers as was the case among the more poetic races of the Mediterranean shores in classic times. The coral is the stony secretion of certain marine polyps, or animals of peculiar organization, in whom a multitude of very simply-formed individuals are united in a compound life, a solid *centrum* forming the means of association. The coral polyp is a marine animal, but the coral itself is a secreted mineral, and its beautiful plant-like appearance is purely imitative, and is no more connected with plant life than the peculiar forms of some insects that may be mistaken for dead sticks and leaves are with true vegetation. Each polyp, however, is possessed of extensile arms or feelers, which almost resemble flowers when expanded,

MINERALS OF ORGANIC ORIGIN. 181

so that there was every excuse in regarding them as
belonging to the vegetable kingdom.

The coral animal that builds islands in warm seas,
constructs reefs of vast extent, rising in vertical walls
apparently from great depths on some shores, and
forms in tropical seas curious circular reefs, like the
fairy rings on a field, which at first might be thought

FRINGING CORAL REEF.

independent of land, is somewhat different from that
which produces the red coral of the Mediterranean
used for ornamental purposes. The result of one
form of those constructions is represented in our
illustration representing an island of circular

CORAL ISLAND AND LAGOON.

shape entirely composed of this material. In these and similar cases the corallium, or hard part, is cellular and almost spongy, it being secreted within the body of the polyp. In the red coral it is secreted by the outer surface of the animal, although it supports the soft parts of the body and is perfectly solid. It is attached to the ground by a foot which is so closely fitted to the rock on which it is found as to be very difficult to detach. There are no roots, however, and this foot does not contribute to the growth of the coral, but from it as a base a stem rises, usually single, and seldom so much as an inch in diameter. From the stem a small number of branches proceed irregularly in various directions. Each of these is studded over with little cells or openings, in which lives a polyp. Unlike the branches of a tree, these coral branches shoot downward towards the bottom of the sea, and thus the appearance of the animal in a living state is less that of a plant than might be supposed from the description or from the specimens when brought up into the air. The coral is generally of a fine red colour, but is occasionally flesh-coloured, yellow, or even white.

The mode adopted to obtain the coral is a kind of fishing performed by divers. Eight men of this kind equip a small boat and carry with them a large wooden cross, with long equal arms very strong, and each bearing a stout bag net. A strong rope is attached to the middle of the cross, and a load being fastened to the centre to sink it, it is let down horizontally to the sea bottom. A diver follows the cross, pushing one arm of it after another into the hollows of the rocks, so as to entangle the coral in the nets, and owing to the inverted position of the branches this is comparatively easy. From time to time the cross and its accompaniments are drawn up to the boat and the diver is changed.

Coral has long been esteemed in the East as well as in Europe, and is obtained from the sea surrounding many parts of Asia. Beads for necklaces are and long have been a form into which the coral was worked, but in India not only necklaces and bracelets but other ornaments have been made from it. It varies much in value according to fashion. In the

East coral is sometimes sold, if the beads are large, for an equal weight of silver.

It is said that an enormous piece of coral was dredged up near Toza, in Japan, in 1879. There were five branches; the stem is 15 inches in circumference and 5 feet in length.

The following description of the coral, though in some points it repeats what has been already quoted in the extract from Orpheus, gives the account of the value assumed to belong to it as a precious stone in ancient times. It is from the poem by Marbodus, already so often referred to:—

> Whilst rooted 'neath the waves the Coral grows,
> Like a green bush its waving foliage shows:
> Torn off by nets, or by the iron mown.
> Touched by the air it hardens into stone;
> Now a bright red, before a grassy green,
> And like a little branch its form is seen;
> Of measure small, scarce half a foot in size,
> A useful ornament the branch supplies.
> Wond'rous its power, so Zoroaster sings,
> And to the wearer sure protection brings.
> Its numerous virtues Metrodorus's sage
> Has told to mankind in his learned page:
> How, lest they harm ship, land, or house, it binds
> The scorching lightning and the furious winds.
> Sprinkled 'mid climbing vines or olives' rows,
> Or with the seed the patient rustic sows,
> 'Twill from thy crops avert the arrowy hail,
> And with abundance bless the smiling vale.
> Far from thy couch 'twill chase the shades of hell,
> Or monster summoned by Thessalian spell;
> Give happy opening, and successful end,
> And calm the tortures that thy entrails rend.

CHAPTER VII.

USEFUL NON-METALLIC MINERALS.

BESIDES the non-metallic minerals already described, there are others of great interest and considerable use, some of them employed occasionally for ornamental and decorative purposes, but which could not fairly be introduced in the preceding chapters. A few of these we may here briefly describe. Other minerals of this group exist, but will not be alluded to in the present work as hardly possessing sufficient general interest to justify a reference to them.

Graphite.

This remarkable mineral, known by various names of which *black-lead* and *plumbago* are the most familiar, is particularly interesting in its relation to the rare, brilliant, and costly diamond, and the much more valuable and abundant coal, on whose presence and convenient position for extraction the national wealth and progress of England have so much depended. Chemically, graphite, diamond, and coal may be said to be identical. They differ only in the mode of aggregation of the atoms of carbon of which each alike is made up. Both diamond and graphite occur in crystals, the former always, the latter occasionally;

but the form of crystallisation is distinct, the latter mineral being generally found when crystalline in flat, six-sided tables, the former in twelve-sided solids, almost globular. The physical characteristics of carbon in the three forms in which it occurs are exceedingly different, and the differences are very remarkable.

Graphite is of a peculiar iron-grey or dark steel-grey colour with metallic lustre. It is absolutely opaque, and feels greasy to the touch. One of its most valuable uses is as an anti-friction material, substances coated with it slipping easily over each other, without becoming heated by friction. Another use is to cover iron surfaces and check the tendency to rust. It can readily be cut, and when pure is too soft to turn the edge of a knife. Used for writing on paper it makes a clean, distinct black mark, which can be rubbed off with bread-crumbs or india-rubber (caoutchouc), and this property renders it useful to the artist and draughtsman. The name graphite (derived from the Greek, γράφω, *I write*) indicates this peculiar property, and the principal use of the purest and best lumps that are found is for manufacture into black-lead pencils. The name *black-lead*, though very familiar, is unfortunate, as the mineral has no atom of lead in its composition.

Although consisting of carbon to the extent of nearly ninety per cent. in most cases, and more than ninety-five per cent. in others, there are generally some foreign substances present in graphite. These consist chiefly of iron and silica, but they only so far affect the mineral as to diminish its value as a lubricator and for drawing. purposes. For lubrication all that is needed is the powder; and this can be purified by washing, at least to some extent, after grinding, but for drawing only the very purest and finest specimens can be used, except for very ordinary purposes. Cheap pencils have sometimes been made by cementing the powder with sulphur, and in the absence of the natural lumps the fine powder alone has been made into material for good pencils by enormous pressure.

The finest lumps of graphite were formerly obtained almost entirely from Cumberland, where they were found in the slate rocks of Borrowdale. Nearly a century ago one of these lumps was discovered which yielded as much as 70,000 lb. of good mineral whose value varied from 30s. to 45s. per lb. This supply was nearly exhausted about the year 1840, when the very ingenious process of compressing the powder into a solid was introduced by Mr. Brockedon, and much of what was regarded as waste was utilized. Soon afterwards, however, a very valuable deposit was discovered in Siberia, and in the Great Exhibition of 1851 enormous lumps of the mineral were shown in the Russian exhibit. The quantity supplied from that source is still sufficient to supply the trade. Mexico, Ceylon, some parts of Germany, France, and Spain all yield a supply of ordinary graphite, and from some districts of the United States and Canada there have been sent considerable quantities, though little of it is of the finest quality.

Ceylon supplies an almost indefinite quantity of graphite of second quality, useful for lubrication and for brightening iron, for which latter purpose it is much used both for household purposes and for machinery. The mineral from Ceylon is generally crystalline in small plates or spangles, and is difficult to convert into use for pencils, even if very pure, as it hardly solidifies, even under enormous pressure.

Inferior graphite or plumbago is sufficiently abundant, and is widely spread. It occurs in small nests or pockets in such rocks as slate, or in rocks of volcanic origin. The value is not great in comparison with that of the pure specimens already described.

Under the name *Plumbago* graphite is largely employed in the composition of crucibles required for smelting steel and in other cases when it is necessary to resist enormous heat. Its effect on the clay which forms the principal material of such vessels is not only to enable them to sustain immense heat, but to give them greater tenacity and expansibility.

Although the Russian exhibition of minerals in the year 1851, already alluded to, contained many ornamental objects cut out of the black-lead obtained from the Siberian mines then recently opened, there has never been any real attempt to employ this mineral for ornamental purposes. Its softness and brittleness, and the impossibility of touching it without a mark being left from it, render it unfit for such purposes. It is eminently a useful mineral, and is employed in the state in which it is found, requiring no preparation except occasional washing to remove grit, and sometimes compression to bring back the fine powder to a compact state.

Plumbago has been sometimes artificially produced in the beds of blast furnaces, and is then called *Kish*. The long-continued heat acting on unconsumed carbon has no doubt produced this result.

There is little doubt that all deposits of graphite are derived originally from either the vegetable or animal kingdom, although they have been so much altered by heat and chemical action as to assume the mineral character. They occur generally in gneiss or other altered rocks quite independently of the presence of coal.

Sulphur.

This rather remarkable mineral, common enough in combination with metals and earths, is rarely found native except in places where volcanic action either recent or extinct can be distinctly traced. It then appears either in lumps or crystals, and, like carbon, the crystallized form belongs to two different systems, the difference being not only in form, but in specific gravity. The principal form assumed is called the acute rhombic octahedron, and in this state the mineral weighs rather more than twice the same volume of water. The other form is obtained when melted sulphur cools rapidly. The resulting crystals weigh something less than twice their volume of water, and are long and needle-shaped.

The true colour of sulphur is pale yellow; but that of Sicily and other volcanic districts is generally of a red tint, due it is supposed to the presence of a little arsenic. In one of the Lipari islands, however, there is an orange-red variety, found near the crater of a volcano, whose colour is due to the presence of a rare earthy mineral called *selenium*, which has several curious resemblances to sulphur, but which is not otherwise remarkable except for its odour, which is that of the horse-radish.

Sulphur melts easily at a temperature not much above that of boiling water, when it becomes perfectly fluid. If, however, the heat is now gradually raised, the sulphur being kept in a closed vessel, the mineral becomes thicker and thicker, till between 430° and 480° Fahrenheit it is so tenacious that the vessel containing it may be inverted for a moment without losing any of its contents; and in this state, if cooled suddenly by pouring it into cold water, it will remain for many hours perfectly soft and flexible, and may be drawn out into threads. It now presents none of the appearances of sulphur, and by exposure and time it becomes brittle and crystalline, although it may at any time be restored to its original condition by remelting and slow cooling. If, however, the temperature be increased above 480° until it reaches at last the boiling point of the mineral, which is 792° F., the sulphur becomes once more perfectly liquid. Heated in the open air whence it can obtain oxygen, sulphur soon takes fire, and burns with a pale blue flame, giving off large quantities of sulphurous acid gas.

Sulphur is often found in oval or rounded lumps in marly beds where there has been volcanic action, and often with gypsum and salt. It is indeed in this condition that most of the sulphur of commerce is obtained; but the relation to its volcanic origin is evident, as the island of Sicily is the seat of almost all the productive mines. The annual produce of that island amounted to about 160,000 tons some years ago, but is now probably less. The mineral is found as a flowery deposit or sublimation about the craters of volcanoes, and in the crevices by which there is communication to the interior. The phenomena called solfateras are of this nature.

Sulphur has a slight odour and hardly any taste, and is almost insoluble in water. When held in a warm hand it emits a crackling sound, and will sometimes break in pieces. This arises from the very slow rate at which it conducts heat. It is also a bad conductor of electricity, with which it is soon strongly charged by friction on a flannel.

Sulphur is used in medicine, but is chiefly required in the composition of gunpowder. To a small extent it is employed as a cement, and for making casts of gems and seals. For this latter purpose it is kept heated a long time, and then assumes the colour of bronze.

Though rarely found in quantities sufficiently large to be of economical value as a native mineral, there are few substances in nature more widely distributed than sulphur. It combines very readily with several of the metals, of which iron, copper, lead, and zinc, arsenic, mercury, and antimony, are instances, and exists as sulphides of these metals in all mining districts. The sulphides of copper, lead, zinc, mercury, and antimony are the principal ores of all those metals, and the sulphide of iron is found everywhere. As a sulphate, or a combination of sulphur and oxygen with various earthy oxides, it is also very common. Sulphate of magnesia exists largely in the ocean; sulphate of lime or gypsum is a common mineral, and sulphate of soda is found in many mineral waters.

Sulphur also exists as a part of the composition of animal tissue and is present in some vegetables. In all volcanic regions where there is communication with the interior of the earth, sulphurous emanations and sulphur salts are exceedingly common.

Rock-Salt.

SALT MINES OF WIELICZKA.

Under the general name of salt, chemists include a vast variety of substances obtained by the neutralizing effect of alkalis on acids; but in mineralogy the name applies only to a small number of species, or rather it is limited to one only, which has no relation to acids and alkalis, being a combination of two elementary substances—the gas *chlorine*, and the metal *sodium*. These form ROCK-SALT. Both the elements are very widely diffused, not only in the earth, but throughout the whole of our solar system, and even in a large number of those stars whose composition has been discovered by the aid of the spectroscope. We only know of the result of their combination as we find it on the earth and in the sea.

Rock-salt is dissolved in the water of the ocean to the extent of four ounces in every gallon, and very large quantities are obtained by evaporation from this source; but it also exists in enormous masses in many countries, and can be obtained by mining operations. Thus in Galicia there are extensive mines which have been long worked, and are now about 860 feet deep. The picturesque appearance of the interior of such a mine is shown in the annexed engraving. At various places along the line of the Alps, and at various elevations above the sea, reaching to upwards of 7,000 feet, there are very large quantities of salt, mixed with, or buried beneath the soil. At Cardona in Spain, near Barcelona, there is a mountain of salt, excavated in steps in the open air, and worked like a stone or slate quarry. In England, especially in Cheshire and Worcestershire, there are magnificent deposits in the red sandstone of those counties. There are known to be two such masses, shaped like a lens, about 32 feet apart, each about a mile-and-a-half long, three-quarters of a mile broad, and 100 feet thick in the middle. These are nowhere much less than 75 feet from the surface; but in some countries where the air is always dry, as in the high plains of Asia and Africa, there are extensive wastes permanently covered with this mineral exposed to the light of day. In Persia, near the Caspian Sea, and in the steppes in the south of Russia, the whole of the soil is impregnated with salt.

The beds of rock-salt are sometimes so thick, that although subject to mining operations on a large scale for many centuries they have not yet been sunk through. Generally, however, the deposits are of moderate thickness, and frequently repeated, alternating with gypsum and often with clays of variegated colours. Besides those places where the salt is reached in a solid form, there are in most countries brine-springs where water, more or less impregnated with salt, wells up to the surface, or is reached by boring. At many such places salt is prepared for the market by evaporation, but much care is needed to separate it from the bitter salts of magnesia, abundant in the water of the ocean.

Salt generally crystallizes in cubes, or modifications of the cube (see fig. p. 17), and when pure is quite transparent, but it is often found tinged with a red or purple colour, owing to the presence of a small quantity of iron. At Wieliczka it is sometimes absolutely pure, but is more often found containing a little of the salts of magnesia and soda. The crystals when pure and transparent possess the remarkable property of allowing rays of heat to pass through them almost without interruption. Thus, when exposed to the rays of the sun, or the intense glowing heat of a furnace, transparent slices of salt do not readily become warm. The difference in this respect between a sheet of rock-salt, one of plate-glass and one of clear ice, all being perfectly transparent to light and of the same thickness and size, is that of 100 rays of heat passed into them 92 pass through the salt, 24 through the glass, and none through the ice. Expressed in other words, only 8 of the rays would remain to warm the salt, 76 would heat the glass, and the whole number would melt the ice.

It is said that in the desert of Caramania salt is so abundant and the climate so dry, that houses are built of this material. At Lahore, in the Punjab, in India, there is a quarry of salt, and dishes, plates, and stands for lamps are cut out of the crystals. Although salt in the ordinary state is so easily acted on by water as to become damp on exposure to the air in England, this is not the case in other countries, and when in the state of crystal and perfectly pure, salt will remain unchanged for an indefinite time. Even in England the crystals will remain dry in places where prepared salt in powder moistens by exposure. The moisture is attracted from the atmosphere.

Salt is one of the few mineral substances entering directly into the composition of food in sensible quantities, and it is regarded as indispensable to health. The celebrated traveller, Mungo Park, says that in the interior of Africa "the greatest of all luxuries is salt. It would appear strange to Europeans to see a child suck a piece of rock-salt as if it were sugar. This, however, I have frequently seen; although in the inland parts of Africa the poorer class of inhabitants are so very rarely indulged with this precious article, that to say a man eats salt with his victuals is the same as saying that he is a rich man. I have myself suffered great inconvenience from the scarcity of this article. The long use of vegetable food creates so painful a longing for salt that no words can sufficiently describe it."—(Park's "Travels," i. 280.)

One of the most remarkable instances of the presence of salt in very large quantities in water is to be found in the Dead Sea, also called the Salt Sea. This remarkable lake, whose level is more than 1300 feet below the surface of the Mediterranean, occupies a singular depression into which the waters of the river Jordan enter and are lost. It is much the deepest surface of water known on the earth. The depth of the water in most parts of the lake near the middle exceeds 1,000 feet, and has been said to reach 2,000 feet in some parts. The mountains that form the wall of this great fissure rise very steeply, and some idea may be formed of the peculiar scenery by the annexed cut, which represents a part of the valley enclosing the lake, where copious brine-springs occur, and where a supposed pillar of salt

USEFUL NON-METALLIC MINERALS. 197

exists. The town has been supposed to be the remains of some buildings in ancient Sodom, but there is no evidence to support the supposition. However this may be, the hills called Jebel Usdum, represented in the view, ranging for seven miles from north to

SALT AT JEBEL USDUM, ON THE DEAD SEA.

south, and having an average elevation of 300 feet, are said to be "composed of a solid mass of rock-salt. The top and sides are covered with a thick coating of marl, gypsum, and gravel, probably the remains of the post-tertiary deposit lifted upon the salt. The declivities of the range are steep and rugged, pierced with huge caverns, and the summit shows a serried line of sharp peaks. The salt is of a greenish-white colour, with lines of cleavage as if stratified, and its base reaches far beneath the present surface." (Kitto's "Biblical Literature," 3rd ed., vol. iii. p. 797). The quantity of common salt dissolved in each gallon of the waters of the Dead Sea is at least one pound avoirdupois, or more than four times that of the ocean, but varies greatly according to the place whence it is taken, the time of year, and the depth. Near the Jordan, and during the rainy season, the diminution in the proportion of salt is very great. The waters at the bottom must be saturated, as large crystals of salt may be dredged up.

The waters of the Dead Sea are not only remarkable for the quantity of common salt which they contain, they also hold in solution a still larger per-centage of chloride of magnesium, and a large quantity of chloride of calcium. Altogether, the solids contained in a gallon of water from this sea weigh about three pounds, or nearly a third of the weight of the same quantity of pure water. At some seasons more than a fourth part by weight of the water of the whole sea consists of dissolved solids. At present the water entering from the Jordan and from the surrounding hills during the rainy season nearly counterbalances the evaporation. Should the rainfall of the district diminish sensibly, the lake would soon be dried up and its bed would be a vast mass of salts, of which common salt would be an important, though not the chief part.

Salts of Soda and Potash.

Nitrate of Soda has some resemblance to saltpetre. Its taste is cool and bitter; it is crystalline and transparent, of a white or greyish and yellowish-white colour, and is often found efflorescent. It is soft, and rather heavier than water, nitre being lighter than water. One great use of this mineral is in agriculture as a manure, increasing the yield of all grass and corn crops in a very striking manner. It is not adapted for use in the manufacture of gunpowder, owing to the readiness with which it absorbs water from the air and becomes damp.

Large and valuable deposits of this mineral are found on the west coast of South America, in Peru, Bolivia, and Chili, at distances varying from ten to forty miles from the coast, and at a level varying from 2,000 to 3,000 feet above the sea. Some of the pampas are covered with a bed of it to a depth of several feet, over an extent of more than 120 square miles. The great desert of Atacama is partly covered with nitrate of soda. It is a stretch of almost uninhabitable country, consisting of wide plains, out of which rise a few ridges and some immense rounded knolls, but there is no river, and no rain falls within its boundary.

Besides being used in agriculture this mineral is economically employed in the manufacture of nitric acid.

Natron, a substance mentioned in Scripture, and there translated nitre, is a mixed carbonate and sulphate of soda common in the East and in Egypt.

There are six natron lakes in a barren valley about thirty miles west from the delta of the Nile, found in a thick incrustation at the bottom of pools evaporated dry in summer. It is used, combined with oil, to make a kind of soap, but it is also employed alone for cleansing purposes and in cookery, either to assist in boiling meat or to commence fermentation.

Nitre, or Saltpetre, is nitrate of potash; it is so called from the name of a place in Egypt where this mineral is found in considerable quantities. It forms into transparent crystals, often needle-shaped, and of a greenish or yellowish-white colour. It has a distinct taste, rather saline and cooling, and is much used in the arts and in medicine. The chief employment is in the manufacture of gunpowder. It dissolves readily in water, producing great cold. Placed on hot coals it bursts into a bright flame; mixed with charcoal and a little sulphur it detonates in the fire, exploding with great violence if confined within narrow limits.

Saltpetre is abundant in many places, especially in warm climates, where organic matter has undergone decomposition. It is found also at the bottom of caverns, as described in the following account given by Kirwan in his "Mineralogy":—"The most celebrated discovery of native nitre was made by Abbé Fortis, in Apulia, in the Pulo, or cavity of Molfetta. In this hollow, which is about 100 feet deep, there are several natural grottoes, in the interior part of which, between strata of compact limestone, nitre is found irregularly crystallized. The stone itself is so richly impregnated with it that it bursts it in many places and forms white efflorescences and crusts, resembling Canary sugar mixed with gypsum on its surface; when these efflorescences are scraped off, more is generated in the space of about a month, but more quickly in summer than in winter."

The chief sources of nitre are in India, Spain, Persia, China, Egypt, Arabia, and South America, but the mineral is also readily obtained in various parts of Europe and in the United States.

Borax, or Borate of Soda, is a mineral found in large lumps on the borders of lakes in Thibet, being obtained by the evaporation of the water of lakes loaded with this salt. It is whitish-coloured, tinged with blue, green, or grey, and is sometimes transparent, but more frequently translucent or opaque. From Thibet it is carried to the East Indies, where it is purified and exported under the name of *Tincal*. It is largely used as a flux in some processes of metallurgy, in the process of soldering lead-pipes, &c., and in the manufacture of a peculiar kind of glass to make artificial gems.

Large quantities of borax are now obtained from hot vapours rising from cracks in the soil in a valley near Tuscany, near the old Etruscan city of Volterra, where the evaporation of the water of lagoons containing the mineral is effected by the natural heat of the water issuing from the earth. In Persia also the water of certain springs is similarly utilized. The same thing occurs in Peru, in Ceylon, and in Canada, the association being always the same, and the vicinity either of recent or extinct volcanic eruptions clearly indicated.

Sassoline is the name of a mineral consisting of a combination of boracic acid and water, with some impurities. It forms small, white, pearly scales, having a soapy feel. It is rather abundant, forming a layer on the top of sulphur emitted from fumaroles in the crater of Vulcano, one of the Lipari islands. *Boracite* is a borate of magnesia; and there are some other curious salts of borax, all more or less closely related to volcanic phenomena.

The process of obtaining from the vapours arising from a volcanic soil the minute proportion (only 0·3 per cent.) of boracic acid it contains, and recovering it by the aid of the heat of the volcanic vapour itself, is singularly ingenious, and obtained a council medal at the Great Exhibition of 1851. The quantity produced was then larger from this process than from the old deposits in the East. Since that time, however, extraordinary quantities have been found in California, near the Pacific coast. The waters of a lake, called Borax Lake, are found to contain no less than 535 grains of borax in the gallon. A similar lake has been described by Dr. T. Sterry Hunt as existing in Canada.

Salts of Baryta and Strontian.

Baryta is an earthy substance derived from the metal Barium, which is never present in the metallic state on the earth. It occurs as *Witherite*, or Carbonate of Baryta and *Heavy Spar*, also called *Bologna Spar*, which is a sulphate of baryta. The name barium is derived from the Greek βαρύς, heavy, owing to its high specific gravity, and the extraordinary weight of its salts is characteristic.

Witherite is common in the lead mines of Cumberland, Derbyshire, and elsewhere in England. It is a hard, brittle spar, used in making fireworks, on account of the beautiful red colours it emits when burnt with saltpetre. It is poisonous, but has no taste. Its crystals resemble those of quartz. Besides its use in pyrotechny, it is employed in the manufacture of glass and porcelain, and is exported to France and Germany to be used in preparing beet-root sugar.

Heavy Spar, or *Barytes*, is found in large and handsome crystals, often detached, and sometimes weighing nearly a hundredweight. It is extremely heavy, of a yellowish-white colour, and sometimes transparent or nearly so, but more frequently opaque. It is found in mineral veins, chiefly in lead-ores, in Derbyshire, but also with those of iron, and in beds with other minerals, and is very widely spread. The white varieties are ground into powder, and used instead of white-lead as a pigment, upwards of ten thousand tons being disposed of in a year. Another use is to make green fire in pyrotechnics, but for this purpose it is combined with sulphur, arsenic, charcoal, and chlorate of potash. When burnt and reduced to powder it becomes phosphorescent.

The metal Strontium greatly resembles barium, but is less heavy, and the earth, *Strontia*, occupies the same relative position as baryta. Its salts, the carbonate and sulphate, are also used for producing coloured fire in fireworks. The colour of the carbonate is pale green, and the mineral is translucent and brittle It is generally found with galena in lead mines. The sulphate, called *Celestine*, occurs in bright pearly-looking crystals, of a pale blue colour, which become phosphorescent when heated, and are very brittle. This mineral is found abundantly in Sicily with sulphur and gypsum, but it is also met with near Bristol, and occurs in France, Spain, Germany, and the United States. Strontia has fewer uses than Baryta.

Lime and its Salts.

Lime is the name given to the earth resulting from the exposure of the metal calcium to the action of oxygen gas. The metal cannot be retained exposed to the air in its ordinary moist state, and is only obtained by the chemist; the earth also, when obtained as lime, absorbs water from the air and soon falls into powder as hydrate of lime. But though neither the metal nor the earth derived from it is to be found or retained in its simple form, the salts of lime are not only permanent, but numerous, and to be found almost everywhere in large quantity. The carbonates are presented as marble, already described, limestone, familiar as a building material, and chalk, too soft for building, but used for many purposes. Mixed with magnesia they form *Dolomite*, a magnesian limestone. The sulphates are alabaster and gypsum, and the fluates occur in the crystals called Blue John. These have been already described. The phosphates are found in large quantity in some countries, and are valuable as the foundation of mineral manures. They occur in

England, in the well-known coprolitic beds in Suffolk, Cambridgeshire, and elsewhere. *Apatite* and *Asparagus Stone* are names given to crystalline phosphates of lime, which are also called *Phosphorite*. They are phosphorescent when in powder. Large crystals of phosphorite have been found in the State of New York, in America; and in New Jersey a shaft has been sunk into a deposit of this substance and the mineral obtained by mining. Masses are brought out occasionally weighing 200 lb.

By far the most important use of the salts of lime is for building purposes, either directly as limestone, or indirectly as mortar, to cause bricks to adhere. There cannot be a doubt that limestone has been employed as a building material from the earliest period; and a large part of the Pyramids, the most

LIMESTONE PYRAMIDS.

ancient buildings of whose age we have historic records, is constructed of this material. The earliest buildings constructed of limestone are also, in some respects, the most remarkable and the most difficult to construct. The old Cyclopean structures of Greece are, in many cases, formed of blocks, each one of which measures more cubic feet and weighs more tons than it would now be found expedient to use for structural purposes, in spite of the many advantages we possess in the way of machinery and steam-power. There is, indeed, nothing more remarkable than the fragments of some of these ancient buildings of limestone, the stone still showing the marks of the chisel and laid in courses admirably finished, which seem as if they could resist for ever the tooth of time. Notwithstanding this, however, the ravages of man have not spared them, and we see in many places how nature has struggled in vain against the barbarous destruction of the most perfect sculptures in this material.

Curious rounded concretions of carbonate of lime are sometimes found such as are represented in the annexed engraving. They have been taken for fossils or for coralline productions, but they are nothing more than imperfectly formed crystals in which true crystalline form fails to be discoverable.

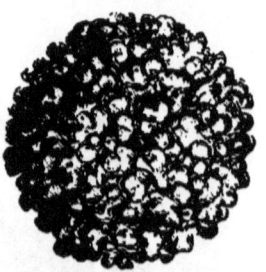

CARBONATE OF LIME, RESEMBLING CORAL.

The varieties of form in which carbonate of lime is found are so numerous as to be impossible to record, even if space were very much more abundant than in nis work. Nor is it necessary that they should be described, as they are all deducible from the same laws and point to the same result. The carbonate of lime is rarely pure, and the substances found with it, whether earthy or metallic, produce effects in accordance with the material and the circumstances under which it is presented.

Magnesia and its Salts.

Magnesia is an earth derived from a metal *Magnesium*, which has become tolerably familiar of late years owing to the intense light given off during its combustion while combining with oxygen gas and becoming converted into the earth. The most interesting mineral connected with it is *Meerschaum*, also called *Magnesite*, a singular white absorbent earth long known and deriving its name, which is German, and signifies *froth of the sea*, from its lightness and white colour.

Meerschaum is an earthy carbonate of magnesia, found chiefly in Natolia, on the plains of Asia Minor. When first dug out of the ground it is soft and soapy, and even forms a lather. It readily absorbs grease, and is employed for cleansing purposes and for washing linen. Its chief use, however, is in the manufacture of tobacco-pipes, for which it is eminently adapted, owing to its porosity and the extent to which it absorbs the oily matter given off from tobacco when smoked. Germany is the chief market for the raw material, and to prepare it for use the bowl of a meerschaum pipe is soaked first in tallow, and then in wax, and afterwards polished.

Besides its absorbent power, which is indicated by the readiness with which it adheres to the moist tongue, it is very smooth to the touch, and so soft as to be scratched by the nail. When broken it shows an earthy fracture.

The Silicates of Magnesia form an interesting series. Some of them, as the Zircon group, have been already described among the gems; others, as Serpentine, among valuable stones. *Hornblende* and *Augite* are found in one form or other among all volcanic rocks. They differ considerably amongst each other in appearance and properties, but with few exceptions require a detailed knowledge of mineralogy to distinguish. There is one exception in the group of which Asbestos is the type.

> Of steely colour and of wondrous might
> Arcadia's hills produce th' *Asbeston* bright;
> For kindled once it no extinction knows
> But with eternal flame unceasing glows :
> Hence with good cause the Greeks Asbeston name
> Because once kindled nought can quench its flame.

Asbestos is one of the most curious of minerals, being almost fibrous, and so little brittle that it is described as mountain-wood, mountain-cork, mountain-leather, mountain-paper, &c., all indicating its peculiar texture, the fibres being interlaced, and more or less laminated or compact. When deprived of foreign substances by exposure to red heat, it may often be separated into loose threads capable of being woven into real cloth or paper, especially when mixed with oil and carefully handled.

Woven in this way it was used by the ancients to wrap the bodies of the dead before placing them on the funeral pile. Thus, the ashes and unconsumed bones were preserved separately for subsequent preservation in vases. Napkins were also made of the same material, which could be cleansed by throwing them into the fire. It is said that certain tribes of Indians make dresses of asbestos, which in a similar way are thrown into the fire to be cleaned. A more permanent use has been found in the preparation of wicks for lamps of various kinds. The ancients used it in this way for maintaining perpetual fire in the temples, and hence the name ἄσβεστος, unconsumable. In Greenland, where the mineral is common, it has long been used for this purpose, and it has lately been introduced in England both for use in gas-stoves, where the appearance of fire is desired, and in wicks for some kinds of lamps. In Corsica, where there is a large quantity obtainable, it is mixed with clay in making a superior kind of pottery less fragile than the ordinary ware.

The whiter and more satin-like varieties of asbestos are called sometimes *Amianthus*. These also have been used for making cloth, and the name is derived also from the Greek (ἀμίαντος, *undefiled*) as expressive of the facility with which it is cleansed.

Alumina and its Salts.

The metal Aluminium, now sufficiently well-known and remarkable for its lightness, being scarcely heavier than glass, bears the same relation to the earth alumina that sodium, potassium, and magnesium do to soda, potash, and magnesia. The earthy substance derived from it by combination with oxygen is, however, even more important, yielding when crystallized some of the hardest stones and most beautiful and valuable gems; and, in combination with silica, an almost infinite variety of minerals, divergent in many essential particulars, but some of them widely spread and in vast abundance The ruby and sapphire have been already described at some length. Corundum has been alluded to, and we proceed to consider other minerals in which alumina holds an essential part. With regard to Corundum, we need only here add that one of its most useful forms is that called *Emery*, a granular variety obtained from Naxos, Smyrna, and other places on the Eastern Mediterranean shores, but common also in Greenland and the East Indies. The precious stone known as Turquoise, which is a phosphate of alumina, has been already described.

Alum-stone, an important salt of alumina and potash, is found native in lumps and well-formed crystals near Rome. It is pearly and translucent. In a massive state it is found also in Hungary, where it is hard enough to be used for millstones. The alum of commerce is chiefly the result of manufacture from a kind of slate called alum-slate, found very abundantly in the Liassic strata near Whitby, in Yorkshire, and in some other places.

Silica, and its combinations with Alumina.

Silica is derived from the metal Silicon. The simple crystalline forms of silica have been described under their better known name QUARTZ. Silica combined with alumina produces *silicate of alumina*, which in its purest state is represented by a pretty crystalline mineral called *Cyanite*, from its light sky-blue colour (Greek, κίανος, blue). It is sometimes spelt *Kyanite*, and is not rare. Occasionally it is transparent, and has been cut and polished for use as a gem, but it is seldom adapted for this. The colour of the blue almost resembles the sapphire, but the mineral is rarely of any special value.

Clay is a silicate of alumina universally distributed, infinitely varied in the nature and proportion of foreign substances or impurities mixed up with it, and exceedingly rare in a pure state. The purest clays are those used in making fine porcelain, and are called Kaolin. Besides common clay and fireclay the material called *Fuller's-earth*, and some others in which water enters largely into the composition, will be found worth special notice.

Kaolin is the Chinese name of certain varieties of clay used in that country in the manufacture of the beautiful ware for which it has so long been celebrated. Similar material is now found in our own country and in many parts of Europe, but chiefly in Cornwall and Devonshire, in England, near Schneeburg, in Saxony, and near Limoges, in France. It is thought to result from a natural washing of the felspar of rotten granite, and in its most perfect state consists of about 45 parts of silica, 37 of alumina, the remainder being partly water and partly lime, magnesia, and potash, with a little oxide of iron and manganese. It varies in colour, adheres to the tongue slightly, but becomes plastic when wet. Freed from the foreign ingredients mechanically mixed with it, the full heat of the furnace fails to produce fusion, and the colour remains unaltered. Like all clay this variety hardens with heat, and more than other varieties acquires condensation and solidity.

Inferior clays of the same kind are called *plastic clay*, but in proportion as they contain foreign ingredients they diminish in value, and are less able to resist the heat of the furnace while retaining their colour. Iron is rarely absent, and the alkaline earths are often present. They act as fluxes, and diminish the value of the mineral.

The quantity of kaolin used in the English potteries is extremely large, upwards of a hundred thousand tons of potter's material being annually conveyed into the Staffordshire and Worcestershire potteries for the purpose of manufacture.

Other clays are valuable for other purposes, and great ingenuity is exercised to obtain plastic material to serve the various requirements of the different trades. *Fire-clay* is needed for fire-brick and crucibles to resist the most intense furnace heat, while a more common clay is employed in the manufacture of the thousands of millions of bricks annually required in our country for constructive purposes. *Pipe-clay*, also, needed for special purposes, must

USEFUL NON-METALLIC MINERALS.   213

not be omitted as one of the varieties of silicate of alumina.

There is no more remarkable or interesting example of the ancient use of brick and pottery earth than that curious monument of antiquity represented in the annexed cut. It is a vast pile of ruins rising to the height of 200 feet, having on its summit a compact mass of brickwork 37 feet in height and 28 feet in breadth. The whole mass is composed

BIRS NIMROUD.

of brick slag and broken pottery, and is called Birs Nimroud, or the Palace of Nimrod. It represents the remains of a building of extremely large dimensions and of great antiquity, having probably been a temple dedicated to Belus, and described by Herodotus.

The building is supposed to have consisted of a series of seven platforms rising one above another, but extending further from the centre in front than behind, so as to be much steeper at the back. The grand entrance was at the back approached by a vestibule, the ruins of which now constitute a separate mound. Similar mounds are known elsewhere, but this is the largest and most important; and the great accumulation of brick and pottery it consists of affords some notion of the vastness of the work undertaken in the construction of Nineveh. The bricks in these cases were cemented with bitumen, and partially burnt.

The ancient Egyptians were great makers of bricks, and they employed their captives taken in war in this manufacture. These, however, were flat white bricks made of a whitish chalky clay, compacted with cut straw and dried in the sun. Such simple materials are well suited to the climate, and were employed both for public and private buildings. Similar bricks are still used. The manufacture was in ancient times a monopoly of the king, the bricks being stamped, and none allowed to make them without authority. The Jews when subject to the Egyptians were treated as captives, and obliged to work at this drudgery. They erected granaries, treasure-cities, and other public buildings, and always of these bricks. In preparing the bricks the clay was brought in baskets from the Nile, tempered and cut into the required shape before drying. Such bricks are often found with the stamp of the king in whose reign they were made, and it is worthy of remark that many more are found bearing the mark of Thothmes III. (the Pharaoh spoken of in the Bible) than of any other king.

Slate is a mechanical modification of clay, little or not at all altered in its chemical composition, and, in fact, generally consisting of such clay as is common on the earth's surface. Ground up with water it is at once re-converted into similar clay. The change that has affected the clay to convert it into slate appears to have been caused by enormous squeezing. By this the atoms of the mineral have been as it were re-arranged, and the resulting solid splits readily enough in planes at right angles to the direction of pressure, and scarcely at all in the planes of the beds in which it was originally deposited.

There are many varieties of slate. The most important is that used for roofing purposes, of which the best kinds are found in Wales, Cumberland, and Cornwall. Some of the Scotch and Irish Slates are also good. The cleaving and dressing of the slates are matters of great nicety.

Whetstone, or *Turkey-stone*, is a peculiar flinty slate of extremely fine grain, and with a large percentage of silica. Its use for hones is owing to this property, and its value depends on the fineness of the grain.

Fuller's-earth is a soft greenish-brown mineral, with a greasy feel, the colour sometimes passing into blue. It consists chiefly of silicate of alumina, with some iron oxide, and about 24 per cent. of water. When placed in water it falls into a pulpy, impalpable mud. It was once used largely by cloth manufacturers in the process of "fulling," helping to cleanse woollen cloth by the absorption of oil and grease of all kinds. At present chemical preparations are employed which are more rapid and certain in their action. It is widely distributed, but the important purposes which it once subserved being otherwise supplied, the extraction has ceased to be a matter of interest.

Allophane is a pale blue, or honey-yellow mineral, translucent like wax, found in small cavities and veins in marl and chalk. It is something like resin in appearance, and consists of silicate of alumina, with upwards of 40 per cent. of water. This and some other minerals have an opalescent appearance, and are curious, but not utilized in the arts.

Felspar is the name given to a characteristic species of a large group of minerals, generally found crystalline, and consisting essentially of a compound silicate of alumina and potash, the potash being more or less replaced occasionally by other alkaline earths and metallic oxides, as soda, magnesia, iron, &c. Many of the minerals of this group are well-known and common, and some have been already described among valuable stones used for ornament. All felspathic rocks are of igneous origin, either directly and evidently, as found near volcanoes and derived from volcanic action, or being parts of rocks that have been so melted or altered. To understand fully this group of minerals requires an extensive acquaintance with the details of geology and chemistry, as well as mineralogy, and we do not propose to give any account of them in this place; at the same time there are some of the group that may be at least referred to. Thus *Pumice-stone*, or volcanic ash, is a light, spongy mineral froth that floats on water. *Obsidian*, or volcanic glass, is a hard glassy mineral, often worked into cutting instruments by savages. It is perfectly black. Amazonstone, moonstone, and Labradorite have been already alluded to in a former chapter.

Before quitting the subject of the natural silicates, so abundant, so varied, and so widely spread, we may with advantage revert to the manufacture of glass, which is an artificial combination of a similar nature, adapted to many uses and serving many important purposes. Glass is a very ancient manufacture, as the beads of the embalmed Egyptians and the heaps of burnt ashes and rubbish, believed by Dr. Schliemann to be remains of ancient Troy, clearly indicate. It is also a very beautiful manufacture, as is indicated by the forms, both chiselled and plain, shown in the annexed cut. Glass is so nearly identical with the volcanic forms of felspar as to render this a proper place to allude to it. We must not, however, be drawn into a discussion concerning a material so full of interest, as we have quite sufficient to do in describing the natural productions, to which our manufactured glass is to some extent due. The subjoined illustration, in showing the forms into which glass can be blown, and indicating the results of cutting and moulding, is a useful conclusion to the present chapter and to this division of the subject, in which an account of earthy minerals has been given. There is hardly any

OBJECTS OF GLASS MANUFACTURE.

practical limit to the variety of form and the beauty of ornamentation of which glass is capable in the hands of a skilful artist.

CHAPTER VIII.

NATIVE METALS.

URAL MOUNTAINS.

WE need not occupy space by describing what is meant by the term metal. Some of the metals are so familiar and common as to be universally known; others, less familiar, resemble them in most of their distinctive characteristics; others are more common, but less valuable; others are difficult to obtain or preserve, and, although unaffected by exposure, are rarely seen. All, however, possess some common peculiarities easily recognisable.

Of the metals a certain number occur in a native state, that is, exist in nature as metals and are reducible by heat alone. These resist the action of oxygen gas in a dry atmosphere, and can generally be retained unaltered. Most of them are valuable, and some of them rare. These are called noble or precious. It is not very easy to say why they are more noble than others, in any ordinary sense of the word; but their rarity gives them an exceptional value. The others, many of them rare, are called base metals. Though sometimes native, they are found generally in the form of earths, from which the metal is obtained by chemical or metallurgical processes.

Of the noble or precious metals gold and silver are the most important and the most valuable. Mercury is a third familiar example. There are also some rare metals of the same group. The native metals not precious are few, but interesting. Several of them are obtained, with other valuable minerals, in the Ural Mountains, and the view of this remarkable mineral district given at the head of the present chapter is intended to communicate a general idea of the scenery whence much of the world's wealth, during the last century and the first half of the present, has been derived. It was only when the great wealth in gold of California and Australia was discovered that the treasure-house of the old world became less considered; but the Ural chain is still specially remarkable and still exceptionally rich in those noble metals found only in a native state.

Gold.

Who is there who is not acquainted with this universal representative of wealth? It has been known from the earliest ages, for in the description of the rivers proceeding from the garden of Eden, in the oldest historic record extant, we read of one that "compasseth the whole land of Havilah, where there is gold." Of Abraham we are told, that "he was rich in cattle, in silver, and in gold." And in the Book of Job, perhaps one of the oldest of the Hebrew Scriptures, we proceed a step farther and learn that "there is a vein for the silver and a place for the gold where they fine it." Elsewhere, in the same book, it is said that "the earth hath dust of gold." At a very early period among the Egyptians, as among the Hebrews, and no doubt also among the Babylonians, Assyrians, Persians, and perhaps throughout the East, beaten gold worked into earrings, plates for overlaying various sacred objects, and head-dresses were manufactured. Among the earliest peoples in all countries gold ornaments were common, and have been found. Dr. Schliemann has exposed treasures of this metal, worked into various utensils, from the ruins of the ancient city he identified with Troy, and similar objects have been met with in almost all parts of the world.

That gold was obtained in large quantities in ancient times is clear from the accounts handed down of the wealth of Nineveh and Babylon, of which the value was equivalent to eleven millions sterling of our money, while the annual tribute paid to Darius, king of Persia, 480 years before Christ, is calculated to have been equivalent to three and a quarter millions sterling. Crœsus, king of Lydia, who lived about half a century earlier, made presents to the temple of Delphi valued at nearly the same amount, of which a large part was in gold. The following curious story, with reference to the treasures of this potentate, is translated from Herodotus, and is suggestive of the mode in which the metal was procured:—"A certain Alcmæon, whose family had offered hospitality to Crœsus, was offered in return as much gold as he could carry away. Providing himself with a large tunic, in which were many folds, he followed a guide to the royal treasury, and there rolling himself among the golden pile, he first stuffed his buskins as full of gold as he possibly could, he then filled all the folds of his robe, his hair, and even his mouth with gold-dust. This done, with extreme difficulty, he staggered from the place, from his swollen mouth, and the projections all around him, resembling anything rather than a man."

As time went on the accumulations of gold increased, and it is said by Appian that the Egyptian treasury, in the time of Ptolemy Philadelphus, contained no less a sum than £178,000,000 sterling. This, however, seems to have been obtained by collecting the whole wealth of the people. The weight of this large quantity would be about 1,250 tons. Enormous as it seems, the quantity was far exceeded in the time of the Roman Empire, and the calculation of a very competent authority shows that the accumulations of metallic wealth (chiefly gold) in Rome alone, at the commencement of the Christian era, must have amounted in sterling value to about £358,000,000. After this there was a great and rapid diminution, the precious metals being, on the one hand, hoarded privately on account of the insecurity of the times, and on the other, spread over a much larger and gradually-increasing area.

Gold is almost always found native and nearly pure, a very large proportion of it being obtained from the sand and gravel of rivers proceeding from mountains, where the gold once existed in veins, and has been removed in course of time by the gradual but incessant waste that takes place from all parts of the earth's surface. In the vein the gold is generally crystalline, occupying spaces in crevices or cavities of quartz-rock, gneiss, and sometimes slate or even limestone. A few very large blocks have been found, and many smaller lumps. These are called *nuggets*, but the great quantity of gold is in much smaller fragments (*pepitas*), or in grains, scales, and fine dust, the latter name being given when the grains are too small to be picked up by the hand. Gold is so heavy that it is easily made to separate from other substances when shaken up with them in water, and in this way the dust is collected. Some specimens of native gold consist of crystals attached to each other, and shooting out in the form of branches of a tree. The crystals are generally regular cubes, or eight or twelve-sided figures.

In ancient times gold was obtained from many countries, and it is perhaps the most widely and equally distributed of all the metals. Although long since abandoned, productive mines once existed and were extensively worked in Nubia and Ethiopia, Africa having been at all times, as it still is, rich in the most valuable mineral productions. Belzoni discovered mining indications in the mountains that surround the great Desert, whence the Pharaohs appear to have drawn treasures of the precious metal, at least six millions sterling being estimated as the annual production; and gold-dust has been from time immemorial a common object of payment for all foreign goods imported from either the Eastern or Western coast. This drain of gold from the interior of Africa has continued unaltered from the earliest periods recorded in history.

Gold-washings, so common in Asia and Africa, were also very familiar even in remote England; Cornwall, Wales, and parts of Scotland having yielded this precious metal to the Romans, besides furnishing the chieftains of ancient Britain with numerous and rich ornaments. Hungary, Sweden, and Norway have all yielded gold, but of the northern countries Siberia has long held the pre-eminence. The works in the Ural were very remarkable, and have been described by Gmelin in the following picturesque manner, which indicates the methods of research :—

"The extent of the works shows that the workmen must have been numerous, whilst an inspection proves that only the first rudiments of the science of mining could have been known to them. Besides some implements, the use of which is unknown, there were wedges and small hammers, all of copper, that had been smelted, but without any particle of gold in them. Instead of large hammers, they seem to have used stones of a long shape, on which are marks showing that handles had been fastened to them. They seem to have scraped out the gold with the fangs of boars, and collected it in leather bags or pockets, some of which have been found. With such imperfect implements the work of excavation must have required the labour of a great number of hands for a long time, and in some cases must have exhausted their patience. In one instance, after having proceeded to some depth and reached a bed of hard stones, the work after penetrating a little way had been abandoned. Some of the pits are 120 feet in depth, shaped like a bell, and are 7 feet in diameter. The passages and props are well executed, but the former are so narrow and so low that it must have been difficult to work in them. The natural pillars left to support the roofs are in some instances still effectual for that purpose, and in these are still found small portions of copper-ore containing particles of gold; in other instances the supports have given way, and in them are found human bones, probably of those who had been buried in the ruins. That a great number of people were employed is inferred from the numerous fragments of earthenware which are found scattered to a great distance around. It appears that only the richest ores were worked, and some of them must have been smelted in the mines, for in the rubbish of one of the supports which had fallen in there had been found melted copper and the implements for smelting it. Some of these implements also have been found on the surface near the pits. The operation of crushing, as well as washing the ores, was performed in the rivulets, and it is supposed that the latter process was omitted in the rich ores, which were found in elevated spots. The smelting, whether in the mines or on the surface, was performed in small furnaces, of which nearly a thousand were observed in the eastern part of Siberia. It may be presumed that a long period must have elapsed since the works were in activity, for the roots of large fir-trees have spread themselves among the stones that are heaped against the sides of the furnaces."

America was discovered in 1492, and the first natives seen wore ornaments of gold, but the quantity brought to Europe for some time was small, and it was not till two centuries afterwards that the rich treasures of Brazil were brought to light. During the seventeenth century, however, the value of the gold obtained was estimated at £337,500,000. Afterwards the yield diminished.

The great discoveries of the present century commenced in the valleys of Sacramento and San Joaquin, near the magnificent bay of San Francisco, in California. In a newspaper published on the spot on the 14th August, 1848, it is said,—" The country for a distance of 120 miles has been explored, and gold found in every part. There are now probably 3,000 people, including Indians, engaged in collecting gold. The amount collected by each man who works ranges from 10 to 350 dollars per day." Such is one of the first notices of the commencement of that great outflow of metallic wealth that has continued ever since to pour forth gold to all the countries of the earth.

A few years later Australia was found to be no less rich than California. The first discovery was in the sand and gravel accumulated in the bends of the Summerhill Creek, feeding a tributary of the Macquarie, but the number of localities soon increased, and extensive mining operations are now rewarded

GOLD-DIGGINGS AT SUMMERHILL CREEK.

by a steady yield, enriching the country and attracting a large population to this distant corner of the world. The annexed view of Summerhill Creek is historically interesting.

By far the largest quantity of gold brought into the market, till within the last quarter of a century, was obtained from surface diggings, varying in depth from a few inches to about 20 feet or more, but since then regular mining operations have been carried on; the quartz, the usual matrix of gold, is now obtained by excavation, crushed by powerful mechanical means, and the fine gold is separated from the quartz-dust by the help of water, often assisted by quicksilver, which dissolves the finest particles of the more valuable metal, reducing them to an amalgam or paste, which can afterwards be reduced by fire, the mercury being driven off and the gold remaining.

Among the most remarkable properties of gold, one is its extreme weight, as compared with an equal volume of any other substance. It is nearly twenty times as heavy as water, and as most of the stony substances found with it are little more than twice the weight of water, and even heavy iron-ores, sometimes found with it, not more than five or six times the weight of water, the difference is great enough to cause an easy separation. Thus, shaking about in water causes a ready separation, and carrying along the fine dust and powder by a current of water, ensures the accumulation of the gold in some convenient spot. In nature, the bends in rivers are often the best places in which to find gold-dust and nuggets.

Another remarkable property of gold is its indestructibility. It is quite unaltered in colour or condition by exposure to any influence, and thus may be preserved uninjured for any length of time. It does not rust away like iron. It does not tarnish like silver. It does not combine with other metals, except in a state of fusion; and it requires a strong heat, about the same as that required by iron, to melt it. Its colour, which is very well known, scarcely varies, except when mixed with copper or silver as alloy, and is even then very easily recognised, whilst although other minerals, such as iron pyrites or mica, are also yellow, the extreme weight of gold makes it readily detected. It is tough, and easily beaten into any shape by the hammer. So remarkable is it in this respect that a single grain of gold may be drawn into a wire 500 feet long, or beaten into leaf which will cover a space nearly 8 inches square. A quarter of a million such leaves laid on each other and pressed together would not form a pile an inch in thickness. A grain of gold is worth about twopence.

Gold is not a hard metal, and to make it fitted for use in jewelry and for coins it is alloyed with silver or copper. Its colour is thus altered a little, but it still takes a very brilliant polish. It mixes readily with these metals when in the fluid state, and it also mixes with iron and with some of the rarer noble metals. Natural combinations with these and with mercury are known, but gold is never found in combination with any of the earthy minerals, with oxygen, sulphur, or carbon. A natural mixture with silver is called *Electrum*. Mixed with palladium, it is found in Brazil and is known as *Jacotinga*. *Auro-tellurite* is a combination with tellurium.

To give some idea of the quantity of gold annually added to the general stock it may be sufficient to state that for the period of nine years, ending in 1872, the annual value of the metal imported into the United Kingdom from all countries averaged eighteen millions sterling.

Silver.

This beautiful metal, though often found native, is more commonly met with in combination with lead. It is also frequently combined with sulphur, and in some important districts with antimony and arsenic. There is another mixture with chlorine.

The exquisite whiteness and purity of colour, the brilliant lustre, the resistance to ordinary atmospheric influences, and the many uses to which it is specially adapted, render silver one of the most interesting and useful of metals. For a long time it was employed almost universally as the common representative of value throughout the civilized world, the dollar introduced by the Spaniards, and weighing about an English ounce (value about 4*s*. 2*d*.), being to this day the common coin in East and West, and recognised alike by the Chinese, the Spaniard, the Indian, and the American. It even reaches the interior of Africa. A nearly corresponding coin is the French five franc piece (4*s*.). The Russian rouble, the Indian rupee, and the German, Dutch, and English florin are all nearly half the value of the ounce of silver.

For some years past, however, the large increase in the quantity of silver produced has so far disturbed the relative value of the two metals that gold has superseded silver as the standard of solid money in most European countries, and the value of the dollar is diminished very sensibly.

Native silver occurs in crystals, in cubes, and in octahedrons, and also in hair-like filaments, branching into beautiful forms, and sometimes massive. It is found in veins in calc spar or quartz, and is generally black when seen in the rock. The best crystals and largest lumps have been obtained from the Norway mines, but very beautiful specimens have been found in several of the Cornish mines, and from France, Saxony, and Bohemia. Peru is exceedingly remarkable for the large quantities of native silver it has yielded, and Chili and Mexico are also very celebrated, but all these dwindle into insignificance in comparison with the recent supplies from the various States included within the Rocky Mountain chain of North-western America.

Native silver is often alloyed with copper, and sometimes with the metal bismuth. It is also found with quicksilver, forming a white amalgam. These are comparatively common in silver-producing countries. *Tellurated Silver* (silver with tellurium) and *Arsenical Silver* are more rare, but the latter comes into use occasionally as an ore. It is easily detected by the garlic odour peculiar to arsenic.

Of the ores of silver the combinations with sulphur are the most characteristic. *Vitreous Silver*, or *Silver Glance*, is a crystalline mineral, containing, when pure, $86\frac{1}{2}$ per cent. of silver. It can be hammered and cut with a knife, and is like one of the ores of copper, but much heavier. It is one of the most abundant of the ores of silver, being found in the principal silver-producing districts. *Brittle Silver Ore*, or *Black Silver*, is a similar ore, in which part of the silver is replaced by antimony, and it is also a common ore. It is less heavy than the former. *Ruby Silver* yields nearly 60 per cent. of silver, and is another sulphide abundant in Mexico, easily distinguished by its brilliant dark red colour. It is sometimes transparent or translucent. There are two varieties, one containing antimony, the other arsenic combined with the silver.

Horn Silver (chloride), when pure, contains 68 to 76 per cent. of silver. It is soft, and cuts like wax or horn, and is an ore much worked in the great mines of Potosi, formerly very celebrated. It is also found in Mexico.

Large lumps of native silver have been occasionally found. At Copenhagen there is a Norwegian lump weighing nearly five cwt. In Saxony a mass was once dug out weighing 168 pounds, and in France lumps of 50 and 60 pounds have been often met with. In China it is said that coins have been struck of native silver.

The production of silver in England for many years has hardly averaged 20 tons per annum, while up to the year 1873 the quantity imported from Bolivia and Chili alone exceeded 8,000 tons, and from other countries amounted to about 4,000 tons. More recently the supplies from North America have greatly increased, and those from South America have fallen off. Mexico, once a chief source of supply, now yields very little. For many years the Isle of Man has had the most productive of the British mines yielding this metal.

A large part of the silver obtained for use is not derived from native metal or silver ore, but is removed artificially from lead manufactured from lead ores. Lead is rarely met with unaccompanied by silver, though silver is often found without lead. The whole of the silver from Great Britain and most of that from North-western America are derived from minerals in which silver and lead are associated.

Mercury.

This is the only metal existing in a fluid state at ordinary atmospheric temperatures on the surface of the earth. It is very heavy, white, and when pure and the surface clean, excessively brilliant. It has the property of forming a pasty mixture, called *amalgam*, with certain metals, and it is used in this form combined with tin-foil to cover the backs of glass mirrors. It has an extremely high reflective power, but on exposure to the air it soon becomes tarnished and covered with a film of oxide. It mixes with remarkable facility with gold, silver, zinc, tin, and bismuth, all of which metals absorb it as blotting-paper sucks in ink. Owing to this property of forming an amalgam, it is used in collecting the exceedingly fine dust of gold obtained by washing gold sands in a running stream or shaking the sands with water in a flat pan. The resulting paste of mercury and gold can be reduced by fire, the mercury passing off into vapour, which can be collected and cooled while the solid gold remains. Mercury is also largely used in the separation of silver in a similar way.

The combination of mercury with sulphur forms a beautiful mineral well known under the name of *cinnabar*. Its colour is a rich bright red, and when worked up into a pigment it becomes *vermilion*. It is common in Europe wherever mercury occurs, and is indeed almost the only ore from which the metal is obtained. The process of reduction is simple distillation. Cinnabar is more or less transparent, and sometimes crystalline.

There are only two localities in Europe where mercury and its ores are at all abundant, one in the Austrian province of Idria, near Trieste, the other in Spain, at Almaden, near the ancient city of Cordova. The yield of mercury was long dependent on the working of these mines, but soon after the discovery of gold in California, large and productive veins of this metal were found and opened, reducing the price almost at once to half that before obtained.

Native Amalgam is a natural combination of mercury with silver, sometimes solid in cubical crystals, and sometimes like a thin paste. It is very heavy as nearly three-fourths of its substance is mercury. It is found in those places where veins of native silver cross veins of mercury.

Mercury is an exceedingly useful metal in chemical science, as it affords the only means of collecting and manipulating numerous gaseous bodies that cannot be collected over water. It is also much used in various philosophical instruments. Its great relative weight and perfect fluidity, combined with the somewhat wide range between freezing and boiling, render it particularly useful for such purposes.

The Platinum Metals.

There is a group of metals which from their mutual association and relation to platinum may be called the Platinum metals. A few words with regard to each of these will be sufficient.

Platinum is heavier than gold, and can only be melted by the intense heat of the voltaic arc. It possesses, with iron, the curious property of welding, that is, the surface can be melted while the mass is solid, and two pieces may thus be made to adhere permanently by hammering. Although only found native, and generally in very small lumps or grains, it is sufficiently abundant. It entirely resists alteration by exposure of any kind, being in this respect fully equal to gold. In thin plates it is easily moulded and flexible.

Owing to its peculiar properties, and to the fact that it does not amalgamate with mercury and is affected by no acid except hydrochloric, platinum is particularly useful for various purposes in chemical manipulation. Its power of resisting heat also makes it available for crucibles to smelt refractory metals. So important are these uses, that in spite of its great cost, which very much exceeds that of silver, it is extensively used in some manufactures on a large scale. It expands very little by heat, but is easily attacked at high temperatures by carbon, phosphorus, silicon, and other elements.

Platinum grains are found in Russian territories with auriferous sands, and as much as 800 cwt. per annum have been thence obtained. The metal is also found in Brazil, in Peru, in California, in Borneo, and elsewhere, but the whole production from these localities is not a tenth of that from Russia. Peru has yielded some large lumps, one shaped like a turkey's egg, weighing more than 25 oz., but far larger specimens have been found in the Ural Mountains, the largest yet obtained weighing 21lb. These specimens are never quite pure, the platinum being alloyed with several of the common metals and with the rarer metals, palladium, osmium, rhodium, and iridium, with all of which it has striking analogies.

Among the uses of this metal it has been attempted to introduce it as a coin, but the attempt was unsuccessful and has been abandoned. It has been employed successfully to give a peculiar steely lustre to porcelain. It alloys with several metals, but in no case with any important result. With iron, however, it makes a malleable, mixed metal, having much lustre, and with copper it also yields a brilliant alloy. It has been found native to the extent of 10 per cent. in an ore of argentiferous copper from Spain, but there is no other instance at present recorded of the presence of this combination.

Platinum has only been known and its properties described since the middle of the last century, and for a long time it was found impossible to reduce it and render it useful for practical purposes.

Palladium is more abundant than platinum, but less useful. It is superior in lustre when polished, and admits of being hammered. It is somewhat harder, and much heavier than iron, but not heavier than silver. It resists oxidation on exposure at ordinary temperatures, but between certain limits of temperature becomes oxidized. In the hammered state it is capable of absorbing 640 times its volume of hydrogen, but in the state of cast metal only 68 times its volume. With gold it forms a white, brittle alloy. An attempt was made by Dr. Wollaston, its discoverer, to introduce it into use for medals, owing to the brilliant polish it takes and its permanence under exposure.

Rhodium is a white metal, extremely hard and durable, but rare and of no practical importance, except in the manufacture of nibs for metallic pens. It is difficult to fuse, and requires for this purpose a much more intense heat even than platinum. It alloys with zinc and tin, but the alloys, though possessing some curious properties, have not been utilized.

Iridium is another curious metal found with the ores of platinum in the Ural mines, generally mixed with another rare metal called *Osmium*. It is extremely hard, and nearly as heavy as gold, and much more valuable. It is a white, brittle metal, infusible, and unaffected by acids, but it combines readily with carbon. The combination with Osmium is called *Osmiridium* or *Iridosmine*. There is another combination of platinum and iridium, *Platiniridium*. There is also a mineral called *Irite*, a mixed oxide of iridium, osmium, iron and chromium, with some manganese. Osmium is a dark-grey metal, and very heavy, but has not been applied to any useful purpose. It is chiefly found with alluvial gold in California.

Ruthenium is the last of the noble metals found in the metallic state alloying platinum. It is heavy, brittle, and infusible, and much resembles iridium in its properties.

Base Metals occurring Native.

Copper.—This beautiful metal, next in order and value to the precious metals, is often found native in very large blocks. It is occasionally crystallized in cubes and octahedra, and is often combined with silver. Very large deposits of the native metal have been found on the shores of Lake Superior, and smaller masses are met with in most places where the ores of the metal abound. A mass was laid bare in one of the Lake Superior mines, about 40ft. long and weighing 200 tons. It contained about $\frac{3}{10}$ per cent. of silver (about 12 cwt. in the whole lump), but the silver was not present as an alloy, being in visible grains, lumps, or strings. When polished, the silver appeared in large spots like the felspar crystals in granite. Metallic copper is found in some cases disseminated in beds of red sandstone, in nodules occasionally assuming beautiful crystalline forms or shooting out like the branches of a tree. Native copper is found in a great many mines in various parts of the United Kingdom and in Siberia, and also in France and Hungary. In Chili it sometimes contains 7 or 8 per cent. of silver.

Though the natural colour of clean copper is the deep, beautiful red with which all are familiar, the metal is slightly translucent when in extremely thin scales, and the colour transmitted is green. It has a distinct taste and a peculiar smell. It bears exposure to dry air without change, but damp air and acid vapours convert it into a green substance, called *verdigris*. It can be hammered into very thin leaf, and drawn into fine wire, which is very strong, as a thread whose diameter is only a tenth of an inch will support 300 lb. without breaking. It melts only at a very high temperature, and is difficult to reduce to a metallic state from its various ores.

Copper is remarkable for the variety and value of its mixtures with other metals. With zinc it makes brass, which is almost a distinct metal, known and freely used by the ancients long before the separate existence of zinc as a metal was learnt. With tin it makes bronze and bell-metal, each much harder than either of the metals of which it is composed. With nickel it makes German silver, now much used in imitations of silver, and capable of being coated with silver by electricity. Some of its ores, as malachite and azurite (the carbonates), are valuable as ornamental stones, and have been already described.

Copper and bronze were used as metals long before it had been found possible to obtain metallic iron and manufacture it into utensils and weapons. The Egyptians certainly were able to harden bronze, or rather to manufacture a mixed metal of this kind which was equal to iron in hardness, and they appear to have known how also to harden as well as toughen copper. The exact nature of the process and the extent to which it may have succeeded it is not easy now to discover, but it is certain that with weapons and utensils so manufactured they could do much that has since been only effected by iron and steel. They could even sculpture the exceedingly hard syenites of Egypt, which have been the means of handing down the sacred language of the Egyptians.

Supplies of copper have been obtained at different times from various parts of the world, and the various ores are very widely spread. In ancient times the island of Cyprus, and, not long after, our own Cornwall, supplied the Phenicians with this metal. The copper from Cornwall continued to be extracted for many centuries, but has now almost disappeared. Chili, in South America, Cuba, in the West Indies, South Africa, and many parts of Australia have all in turn flooded the market; and Russia, especially from the Ural Mountains and Siberia, has yielded vast quantities. The native copper from Lake Superior has already been mentioned, and many other parts of the United States have been found rich in this metal. There are productive mines in Newfoundland. It

It occurs indeed to a greater or less extent in every principal mining district. The French have always shown themselves particularly clever in manipulating copper, and hammering it into very complicated forms.

Bismuth.—This metal, which is hard, brittle, and of reddish-white colour, is chiefly found native. It is little affected by exposure at ordinary temperatures, but when red-hot oxidizes rapidly in the air. It melts readily, and when becoming solid after being fused expands considerably. Owing to this peculiar property it is useful in the preparation of some alloys. This is especially the case with type-metal, a mixture of antimony, lead, and tin with bismuth, requiring to be hard, but not brittle, and needing to take the form of very minute peculiarities in the mould. The effect of the expansion of the bismuth when solidifying is such as to drive the metal into all these crevices. Another use of bismuth is to reduce the melting-point of some alloys. Thus, a mixture of eight parts of bismuth, five of lead, and three of tin produces a metal that melts at a heat a little less than that of boiling water, although none of the three metals separately would melt at twice that heat. There are no important ores of bismuth, and when found it is generally impure, owing to the presence of other metals, such as iron, lead, silver, arsenic, &c.

Bismuth is remarkable for the peculiar crystals formed when it has been melted and is allowed to cool. These are shown in the accompanying cut.

DISMUTH.

They are not true cubes, and are often hollow or depressed at the surface, and marked in a very striking way.

Bismuth is remarkable in its relations to magnetism. Iron and certain other metals are magnetic, that is, if a bar of iron be hung between the poles of a horse-shoe magnet it arranges itself along the line that unites the two poles. If a bar of bismuth is so suspended, it arranges itself at right angles to this direction, and is thence called diamagnetic.

Arsenic is a crystalline brittle metal, with brilliant lustre, and of steel-grey colour. It occurs native or with antimony, and sometimes with traces of other metals. In its metallic state it does not change on exposure to dry air, and is not poisonous, but when wetted and powdered it oxidizes, and is then an active poison. Heated it gives off fumes smelling strongly of garlic, and these too are poisonous. At a red heat it burns with flame. It may safely be kept unchanged in pure water. Combined with oxygen it becomes *white arsenic*, used in medicine; and with sulphur it forms *orpiment* and *realgar*, used as pigments and for making coloured fire in pyrotechnics. Hungary is the chief source of supply, and in that country and in the Tyrol arsenic is frequently eaten to some extent by human beings and given to horses. The effect when not injurious is to produce sleekness and a fine skin.

The peculiar hardening property of arsenic makes it a useful ingredient in some manufactures, and the colour is valuable in others. It acts as a flux, and also gives an opalescent appearance to glass-ware. It hardens lead in making shot. The yellow and red pigments are very valuable in calico-printing, and the poisonous qualities of the oxide are utilized to kill insects, and thus prevent damage to valuable property.

Arsenic and its poisonous properties have been long known, and its abundance, the facility with which it can be prepared, and the resemblance some of its salts bear to harmless powders, have caused it with some reason to be regarded as a very dangerous substance. It is not, however, without important uses and does not always act unfavourably on the human frame. It may be taken occasionally in small quantities with no danger, but if frequently repeated it appears to act on the system, and produce a greater and more mischievous effect than a larger dose taken at once. There is no doubt, however, that it is commonly used in Styria with or accompanying food, and in that country no ill-effect is proved. There is no evidence as to how far climate and local conditions cause this exemption.

Worn-out horses are by it endowed with new vigour, and pigeons show greater liveliness and increased appetite when under the influence of small doses of this preparation of arsenic.

The appearance of native arsenic is peculiar, and one of its forms is illustrated in the annexed cut. It is often accompanied by red silver ore and blende.

NATIVE ARSENIC.

The mineral is arranged in concentric layers which frequently shell off by successive coats like onions.

Meteoric Iron.

The exceedingly useful and abundant metal so familiar under the name of Iron is only found native in masses which there is reason to suppose have passed through the atmosphere, and belong to bodies circulating round our sun independently, but occasionally coming within the influence of the earth, when the paths of the group of particles to which they belong happen to cross the orbit of the earth at a time when the earth is near enough to attract them. They are called meteors, and the iron is *meteoric iron*. There is no doubt that the number of such bodies is enormous, but the greater proportion are probably dissipated before reaching the earth.

NATIVE METALS.

Meteoric iron almost always contains nickel. It also contains cobalt, chrome, and other metals, and carbon, sulphur, phosphorus, magnesia, and other earths. All these are in small proportions. The iron is capable of being at once worked into utensils, but the lumps are generally covered with an exceedingly thin black crust, the result of fusion, owing to the intense heat produced by friction in passing through our atmosphere with extreme rapidity, and the specimens that have fallen are often buried deeply in the earth owing to the same cause. The metal is in a crystalline state, and when polished and subjected to the action of acids the surface becomes marked with a curious system of lines, from which an impression may be printed on paper.

Masses of meteoric iron of considerable size have often fallen. A Swedish expedition to Greenland about ten years ago brought back a number of specimens, one of which weighed twenty-one tons and another nine tons. The former is in the museum at Stockholm, the latter in Copenhagen. They contain about five per cent. of nickel. They were lying loose on the shore near basaltic rocks. A specimen was obtained about a century ago in South America, weighing fifteen tons, and another in Brazil of about eight tons. Nine-tenths of all the meteorites that have fallen within the last fifty years have been within one of two zones: one in America, ranging N.E. and S.W., about a thousand miles in length, and another in the old world very much longer, and nearly parallel. The breadth of each zone is about 500 miles.

A very complete and interesting collection of meteoric irons will be found in the British Museum.

Iron may be regarded as really the most precious of all metals. It is moderately heavy, takes a brilliant polish, is the hardest of all that are malleable and ductile, and is the most tenacious of all. Its relation to magnetism is exceedingly remarkable, and it is singularly susceptible to the magnetic current when exposed to it in a pure and soft state. Soft iron soon, however, loses all traces of magnetic power when the influencing current is withdrawn. It can be wrought into any required form, drawn into wires of any strength or fineness, or rolled into plates or sheets of almost any thickness or thinness, from 10 or 12 feet thick to the thinness of a sheet of fine paper. It can be bent in any direction, and at pleasure can be stiffened. By admixture with a little carbon it can be converted into steel, and is then capable of being rendered almost as hard as the diamond. When hard it retains permanently the magnetic power. By another mixture with carbon it can be rendered so fluid when melted as to penetrate the most minute crevices of a mould. "It accommodates itself to all our wants, our desires, and even our caprices. It is equally serviceable to the arts, the sciences, agriculture, and war ; the same ore furnishes the sword, the ploughshare, the scythe, the pruning-hook, the needle, the graver, the spring of a watch or carriage, the chisel, the chain, the anchor, the compass, the cannon, and the bomb."[1]

[1] Ure's "Dictionary of the Arts," 7th edition, p. 918.

Steel, as mentioned in the last paragraph, is a modification of iron. Formerly all valuable steel was obtained by treating the finest kinds of iron in crucibles with carbon, and was called cast-steel. When required for the finest purposes of cutlery and philosophic instruments this method is still followed. But enormous quantities of steel for more ordinary purposes are now made directly from iron while in the process of manufacture by forcing atmospheric air through cast-iron in large quantity. The methods introduced by Sir H. Bessemer and Dr. Siemens for procuring steel in large quantity at moderate cost are extremely ingenious, and will no doubt lead to the adoption of this material in a large number of cases, where its cost would before have excluded it. It is already taking the place of iron rails, and has begun to be used for shipbuilding. The much smaller thickness, and consequent diminished weight of steel when compared with iron, in cases where a certain specified strength or resisting power is required, renders it often not only better, but even cheaper, as a material.

CHAPTER IX.

Ores or Minerals yielding Common and Useful Metals.

Most of the really useful metals—by which is meant the metals employed in manufacture, and not merely those valuable for their beauty, their use as coins in representing labour, or their rarity—are obtained from minerals called *ores*, which are sometimes metallic in appearance though frequently earthy. These ores are buried in the earth, sometimes regularly deposited in beds, but more generally they occupy fissures or cavities in rocks. In the former case they are reached with comparative certainty, and the mineral extracted easily, but in the latter a knowledge of mining, obtained by experience, is needed in the search, and often much ingenuity and trouble are required before the ores can be found, and many processes carried on before they are prepared for the chemist and metallurgist, by whom they are to be reduced to the metallic state. Some of the ores are, however, useful before becoming metals, and pass into the market in this state.

The object of the present chapter is to give an account of the minerals from which are obtained such metals as are most widely known and are used for economic purposes. The number of the metalliferous minerals is exceedingly great if we include all varieties that might be so employed, but if limited to those which are sufficiently plentiful to justify their use on a large scale it dwindles down at once to a comparatively small list. There are for most of the useful metals certain ores widely distributed and generally abundant, but only common in one form at any one place. Thus we get iron from one class of ores in one part of England, and from another class at another part. From Sweden we get magnetic ores, and from Elba specular iron. In Yorkshire excellent iron is made from the carbonate; in Lancashire, iron equally good or even more excellent from hæmatite. It is true that the qualities of iron differ very greatly, but this is not the case with all metals. The lead made from the sulphide differs in no respect from that made from the carbonate. In the case of zinc, also, the zinc from calamine is identical with that from blende.

It will, therefore, be necessary to consider the groups separately, with a view partly to the nature of the ore and partly to the quality of the metal. This requires some special considerations with regard to each. Besides the ores actually and largely used to obtain the metals there are a few ores valuable for other purposes, and in which the metal, though predominant, is only a product of secondary importance.

Ores of Iron.

For practical purposes, iron is always obtained from natural combinations of the metal with oxygen gas, either directly or indirectly. By this is meant that some of the ores are oxides and some carbonates of the oxides. These include all varieties utilized for the metal, but the value and importance of the members of the groups thus formed vary exceedingly. Combinations of iron with sulphur are also exceedingly common, and not without value, but not for iron-making. Combinations with phosphorus are numerous and interesting, and are now likely to be usable, when occurring in the carbonates. There are many other combinations forming minerals of more or less interest, and among them the silicates are the most important, but they can hardly be said to be of use in iron-making.

Magnetic Iron Ore is the richest in iron of all the ores. It consists of $71\frac{3}{4}$ parts of iron to $28\frac{1}{4}$ of oxygen, and is a combination of two oxides (the protoxide and the peroxide). Being one of the combinations nearly always free from foreign admixtures it has great value for making the purest iron. As the name indicates, it is magnetic; small particles are attracted by the magnet, and it attracts small fragments of iron or steel without preparation. *Lodestone* is a name often given to specimens of native magnetic iron ore.

In England, which is so rich in many ores, little of this kind is found, but in Norway, Sweden, and Russia almost all the iron is made from it. Great hills in Sweden are made up of it, and the mere loose masses found at the foot of one of them have served to feed extensive iron-works for nearly 200 years. The island of Elba, near Leghorn, contains it with other ores, but not so abundantly. It is found in many parts of Germany, in France, in Italy, in the East Indies, in Chili, in Brazil, in many parts of the United

States, and in Canada, and also in Africa. There are many localities in England in which it is met with, but nowhere in sufficient quantity to compete with other ores less rich, but more plentiful.

This ore is iron-black in colour, it breaks with a metallic lustre, is brittle, and often found in very beautiful crystals. The Elba specimens are particularly celebrated in this way. It is not uncommon in New Zealand, but occurs there as a magnetic sand.

The natural magnet, or lodestone, is not only interesting as an ore of iron, but in its relation to magnetism. It was early known to possess the remarkable power of attracting iron, and also the property of polarity. Such properties in the Middle Ages were not likely to escape notice, and the supposed powers of the magnet are thus described by the poet so often quoted:

> The magnet gem crowned India brings to light,
> Where lurks in caves the gloomy Troglodyte ;
> Coloured like iron and by nature's law
> Appointed iron itself to draw.
> The sage Deendor, skilled in magic lore,
> First proved in mystic art its sov'reign power.
> Next far-famed Circe, that enchantress dread,
> To help her magic spells invok'd its aid.
> If a sly thief slip through the palace door,
> And strew unseen hot embers on the floor,
> And powder'd lodestone on these embers spread,
> The inmates flee, possess'd with sudden dread :
> Distraught with horrid fear of death they fly,
> While from the square the vapour mounts on high.
> They fly : within the house no soul remains,
> And copious spoils repay the robber's pains.
> The lodestone peace to wrangling couples grants,

> And mutual love in wedded breasts implants :
> It gives the power to argue and to teach :
> Grace to the tongue, persuasion to the speech ;
> The bloated dropsy, taken in mead, it quells,
> And sprinkled over burns their pain dispels.

The magnetic ores have been used from time immemorial, or, at least, since iron entered into use at all. They can be reduced with comparative ease with a blast in small furnaces, and the work as carried on, no doubt thousands of years ago, in the East, is still to be seen practised by the natives in the smaller country districts of India. Till lately all the fine steel of the Damascus and Toledo blades, and all that was most exquisite and valuable in manufactured iron, was obtained from this ore. The almost total absence of sulphur and phosphorus in the magnetic ores rendered it easy to obtain from them without trouble the purest steel.

Steel manufactured from magnetic ores requires but little manipulation. It is obtained by mixing pure iron with a certain amount of carbon, and when this is done the mixture becomes capable of many important uses to which it would otherwise be inapplicable. Magnetic ores, however, though among the best, are not the only ores from which steel is made, nor is this kind of ore essential to the manufacture.

Hæmatite.—When pure this valuable ore contains $69\frac{1}{2}$ parts of iron, the rest being oxygen. There are two kinds, one metalloid, or in a metallic form, bright and crystalline, the other earthy. Both are red, both when reduced to powder and when scratched, but the former in its ordinary state is almost black and shining, while the latter is always blood-coloured. The origin of the name and the presumed properties are thus described in the quaint and exaggerated language of the poet of the Middle Ages, who interested himself in stones, and whose lines have been so often quoted:

> The Hæmatite—named by the Greeks from blood—
> Benignant Nature formed for mortal's good.
> Its styptic virtue many a proof will show
> To heal the tumours that on th'eyelids grow
> And rubbed on darkening eyes it clears away
> The gathering cloud and gives to see the day;
> Rubbed in a mortar with tenacious glaire
> And juice of pomegranates, an eye-salve rare.
> Those who spit blood its healing power will own
> As those who under cankering ulcers groan;
> Dissolved in water 'twill allay the smart
> Of poisonous serpent's bite or aspic's dart.
> If mixed with honey 'tis an unction sure
> All maladies that pain the eyes to cure.
> Of red and rusty hue, in Afric found,
> Or in Arabian, or in Lybian ground.

Most of this reference is to the varieties of hæmatite that show clearly the red colour, and which have a fibrous appearance radiating from a centre, as shown in the accompanying illustration. There are, however,

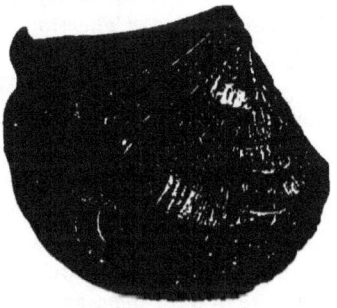

HÆMATITE.

other kinds that must be considered, and that are more metallic in appearance.

Specular Iron Ore is the name given to a variety of hæmatite showing crystalline structure and high metallic lustre. It is especially abundant in the Isle of Elba, where it has been worked for upwards of 2,000 years, and where it occurs in enormous veins or fissures in crystalline rock. The crystals are extremely beautiful, and the rich red colour natural to the mineral is shown in the fracture, and when specimens are scratched or powdered. It is occasionally feebly magnetic. Its colour is often remarkable for a curious iridescent appearance. It is widely distributed, and is a valuable and excellent ore, free from sulphur and phosphorus.

Micaceous Iron is a variety occurring in thin plates or spangles like mica, which separate on touching, and stain the finger red. *Oligist* is the name given to another variety found in Brazil, or rather it is another name for specular iron.

There is a very remarkable deposit of this kind of hæmatite in Missouri, U.S., thus described in Dana's " Mineralogy " :—" The Iron Mountain is 300 feet in height, and consists wholly of massive peroxide of iron lying in loose blocks, which are largest about the summit, many of them from ten to twenty tons in weight. The Pilot Knob is 650 feet in height. It is capped with specular iron, which has the appearance of stratification, and is micaceous in structure." Similar deposits have been observed by myself in the State of Virginia and in North Carolina. The quantity of fine, pure ore in these deposits seems practically unlimited.

Red Hæmatite is the general name of the earthy

and fibrous varieties. *Fibrous Red Iron Ore* is another name for those which most clearly present the fibrous structure. They are found in considerable abundance in Cumberland and Wales, and also in Spain, both in the north and south of Germany, and in Russia. The ore is valuable, and makes excellent iron.

Red Ochre is an earthy form of hæmatite, soft, and soiling the hand or paper. *Reddle* is an impure variety used for marking sheep, and sometimes prepared from iron ores. It is sometimes made into pencils, and is also used as a polishing powder.

Hæmatite takes a high polish when hard, and can then be used for polishing glass, gold, steel, and other metals. It is also remarkably useful for laying on metals before gilding. The finest grain and purest texture of the mineral are required for burnishing, and specimens adapted for this purpose are almost entirely obtained from Galicia in Spain. The powder of hæmatite is very useful in polishing glass and metals, and also as a colouring material.

Brown Hæmatite.—This name includes a number of minerals formed by the combination of the oxides of iron with water, and thence called hydrous oxides. At least a dozen minerals are included. *Limonite*, also called *Bog Iron Ore* and *Brown* or *Yellow Ochre*, are varieties sometimes used as ores of iron. They all belong to the same species, and contain, when pure, 83 per cent. of oxide of iron, the rest being water, but they are apt to be mixed with phosphorus, and in that case are not adapted for use as an ore. Ores of this kind are supposed to be derived from the acid of decayed vegetable matter in swamps, on oxide of iron present in the rocks adjacent, whence the name limonite, which means meadow-ore. Bog-ore explains itself. None of these ores occur in a crystalline form, but they are often in rounded aggregations, like bunches of grapes, or in stalactites, as if formed in caverns. Limonite contains, when pure, about 56 per cent. of iron. Its colour is brown or yellow of various shades.

Yellow ochre is the name of a variety resulting from the decomposition of limonite found in many places, especially in Anglesea. When washed and ground it is used as a pigment, and when also burnt it forms another pigment of reddish-brown colour. *Umber* is an earthy variety containing manganese, found abundantly in the Isle of Man, in the Forest of Dean, in Gloucestershire, and in the island of Cyprus.

Spathic Iron is a crystalline carbonate of the oxide of iron, and is found both crystalline and earthy, and in very large quantities. It is one of the most abundant and valuable of all the ores of iron, nearly all the iron in many parts of Austria and Germany being made from it. In Spain also there are large deposits. Another carbonate, not crystalline, but of the same nature, is the ore commonly used in the British Islands. Besides the two principal divisions of the carbonates into crystalline and earthy, the various earthy ores differ much among each other.

Sparry Iron Ore, as the crystalline carbonate is often called, is of a pale yellowish-brown colour when broken, and darkens on exposure. In Styria there is a vast deposit yielding a quantity practically inexhaustible, but the fuel for reducing it is not at hand. In the Hartz Mountains there are other large veins to which the same remark applies. In Spain, in the province of Biscay, there is a whole hill composed of it, which has been worked for some thousand years, but shows no symptom of exhaustion. In England it is found in Cornwall, Devonshire, and Somersetshire, but not in large quantities. It has, however, been worked to some extent. It affords an iron well-adapted for conversion into steel, and is called by the Germans for this reason *Stahl-stein* (steel ore). Theoretically, it contains 62 per cent. of the protoxide of iron (about 44 per cent. of iron), but except in the crystalline state the quantity of metal is less, and oxide of manganese is not unfrequently found with the oxide of iron even in this case. The crystals of sparry iron are very often in a starry form, radiating from a centre. Hence the names *Siderite*, or Star-stone, and *Sphæro-siderite*, often applied to them.

The *Clay Ironstones* which are very varied in appearance and value, and numerous ores recently brought into use in England, belong more or less directly to this class. For a long time the *Black-band* was the most valuable of these—being very free from sulphur and phosphorus, and containing carbon and limestone, which added to the value of the ores. The common clay ores, less rich in iron and more loaded with impurities than the oxides, long afforded means of making an enormous quantity of useful metal, and for many years England, Wales, and Scotland enjoyed through their use the privilege of supplying the world with cheap iron, to the great advantage of our country. At a somewhat later period the oxides came into larger use, and the quality of the iron was thus improved; and about the same time a process was introduced by which a kind of steel was made direct from these rich and pure ores. This process, originating in the experiments of Sir H. Bessemer, has since been modified by Dr. Siemens and others, and at this time the impurities arising from the presence of phosphorus are expected to be removed by a still further important improvement in the operation of smelting. These modifications apply to all ores, and are likely to enrich other nations as well as England, enabling them to make the best use of their ores.

After the clay ironstones there are few ores valuable for smelting purposes. *Vivianite*, or *Native Prussian Blue*, is a phosphate of iron, somewhat abundant, but of little use. There are several silicates of interest to the mineralogist, but not used in iron-making. There are also sulphates, arseniates, and chromates of iron, showing how widely the metal is spread and how many combinations it is capable of. The chromate of iron is only valuable in the preparation of chrome.

There remains only one more mineral of interest and abundance connected with iron. It is the sulphide or combination of the metal with sulphur. It is called *Iron Pyrites*, or *Marcasite*, and sometimes *Mundic*. Combined also with arsenic it forms *White Iron Pyrites, Arsenical Pyrites*, or *Mispickel;* combined with copper, as it sometimes is in very small proportion, it is called *Cupriferous Iron Pyrites*.

ORES OR METAL-YIELDING MINERALS.

All the minerals called pyrites have common properties from which their name is derived. They contain sulphur as an essential part, and they strike fire with steel. When broken they emit a sulphurous smell. They are hard, and somewhat heavy.

Iron pyrites generally crystallizes in cubes striated on the surface, but with the striæ on adjacent faces at right angles to each other. Specimens are common in almost every copper-mining district, and also in chalk, slate, volcanic rocks, and many veins in old rocks. There are appreciable differences in colour among pyrites indicating the presence of other metals than iron; those which are white containing arsenic, and the varieties which are variegated containing more or less copper.

The great use of pyrites is for obtaining sulphur for the manufacture of various sulphur salts. Enormous quantities of poor cupriferous pyrites are brought from the south of Spain; the value of the exceedingly small percentage of copper (generally less than 3 per cent.) being sufficient to pay for the cost of transport of the mineral and the extraction of the copper, and thus rendering them profitable. Copperas (sulphate of iron), sulphuric acid, and alums are the principal products obtained from the reduction of iron pyrites.

The annexed cut illustrates a peculiarity of stria-

IRON PYRITES.

tion of iron pyrites, and at the same time shows the form in which crystals of this mineral are often grouped. It may be seen that the striæ, or lines, running along parallel to the side of each face, always make right-angles with the lines on the next adjacent face. These striæ are permanent, and if a new face is produced by chipping one away this new face is found to be striated in the same direction. It is not difficult by means of these striæ alone to recognise cubical pyrites.

Ores of Copper.

In the case of copper the sulphides are the more ordinary and abundant ores, the oxides and carbonates though very familiar being rarely found so largely developed as to form the principal ores. Of the sulphides, the richest is called *Vitreous Copper Ore* or *Copper Glance*, and when pure, contains nearly 80 per cent. of copper. It is a blackish-grey looking substance, and melts in the flame of a candle. It is moderately soft, and often found in crystals shaped like the head of a nail. There are several varieties, but none are so abundant as to render this class of ore very important, though it is highly appreciated whenever met with.

Copper pyrites is always a combination of iron and copper with sulphur, but the crystalline form differs from that of iron pyrites. It is a beautiful mineral, resembling gold in its rich yellow colour, which is deeper as the proportion of copper is larger, and valuable in proportion to its softness. It rarely contains more than 30 to 35 per cent. of copper, but though a comparatively poor ore (the copper often passing down to under 3 per cent.) it is on the whole the most valuable of all, owing to its far greater abundance.

All the pyritous ores occasionally contain the precious metals, and both auriferous and argentiferous pyrites are common. The gold or silver is often, perhaps always, present in the native state. It is rarely sufficient in amount to pay for extraction.

Fahlerz, or *Grey Copper Ore*, is a mineral found with other ores, and is very rich. It is a compound of sulphur and copper with antimony, and generally also contains silver. In some ores silver replaces the copper, and produces *Silver Fahlerz*, a very valuable ore.

Red Copper Ore is an oxide often crystalline, sometimes earthy, and contains nearly 89 per cent. of copper. Its deep red colour is very striking, and specimens are sometimes translucent. It is found

RED COPPER.

in most copper-mining districts in the British islands and abroad. It is a very beautiful ore, and crystallizes into forms of which a good idea will be obtained from the annexed cut. When abundant it is one of the most valuable ores, but large quantities are rarely found in any of the copper districts.

Tenorite, or *Black Oxide of Copper*, contains rather less than 80 per cent. of copper, and though sometimes metallic in appearance, it is most usually a black earth. A quantity of it was found some years ago in Tennessee, but it was soon exhausted. It appeared to be produced from the decomposition of other ores. Though somewhat resembling manganese, it is very uniform in its appearance and easily distinguished by the stain it makes when moistened.

Malachite and *Azurite*.—These beautiful ores of copper, both of them varieties of the carbonate, have been already referred to in another chapter, but besides those specimens which can be cut into thin veneers for ornamental purposes, there are large quantities obtained unfitted for these uses. As an ore of copper malachite was at one time confined to Siberia, but it is now common in South Australia, and enters the market for the use of the smelter. When pure, the carbonates contain 70 per cent. of copper, and they are in many cases derived from the decomposition of the sulphides.

There are some silicates of copper containing about 50 per cent. of the metal which enter into use as ores, and the sulphate (*Blue Vitriol*) is sometimes found native though it is more frequently manufactured. There are several phosphates and arseniates of copper forming curious minerals, but of no practical utility.

Ores of Tin.

Tin is a metal very useful in many ways, though

chiefly so in its relations with other metals. It may be beaten into thin plates (tin-foil), especially if worked at the temperature of boiling water. It fuses readily, far more readily than lead, and burns with a bright flame. It is heavy, and has a peculiar taste and smell, the smell being particularly observed when the metal has been long held in the hands. It is very readily acted on by mercury; thin sheets of tinfoil, formed into a kind of paste or amalgam, being used as the reflecting surface in glass mirrors.

There is practically only one ore of tin, the oxide, sometimes called *Cassiterite*. When in the shape of rounded lumps with concentric and radiated structure it is called *Wood Tin*. It is more generally crystalline, the form of the crystals being shown in the accompanying illustration. Tin ore has long been known to exist in association with granite rocks in the extreme western countries of the old world, having been found in Cornwall, Brittany, and the western extremity of Spain. It has been found in some quantity in Australia, but more especially abounds in the island of Banca, where, under Dutch supervision, it has long yielded supplies capable of governing the market. In almost all cases the oxide of tin is either obtained by grinding and washing poor granite veins, or by washing the sand and gravel that have been derived from this granite, and which have been to

CRYSTALS OF TIN OXIDE.

some extent sifted by running water. Such ore is called *Stream Tin*.

Ores of Lead.

Lead is a very useful metal, widely distributed, but almost always either in combination with sulphur, as galena, or as carbonate derived from the sulphide. These ores are so usually associated with limestone that there would seem to be some relation between the two minerals, dissimilar as they are. Occasionally galena is also found in veins in granitic and slaty rocks, and in these cases it is usually of good quality and combined with a certain quantity of silver.

Pure lead is a metal of remarkably close texture, in this respect resembling gold. It is inelastic, malleable, soils the finger when rubbed, and marks paper, and it emits a peculiar odour when rubbed. It is poisonous in many of its combinations, and to a small extent soluble in pure water. It is used in the arts in many ways, both as a metal and in combination. Alone, it is required in sheets for roofing, in pipes for water and gas, and for many other purposes. With tin it makes pewter and solder for fastening metals together; with antimony and tin it makes type-metal. Combined with oxygen it makes one yellowish mixture called *massicot* and *litharge*, used in glass-making, and another of red colour, called *minium* or *red-lead*. The carbonate is *white-lead;* the chromate is used as a yellow pigment; and the acetate is called *sugar of lead*—a curious poison distinctly sweet to the taste and used in some manufactures.

ORES OR METAL-YIELDING MINERALS. 265

Galena, or *Lead Glance*, is the sulphide and the principal ore. It is brittle, brilliant, and often crystalline. The annexed illustration gives an idea of its common form. It is found in many parts of England and Wales, and also in Scotland, and is very abundant in the Isle of Man. It abounds in France, Spain, Germany, and Belgium. It is found abundantly in North America, and is common in many parts of Asia. Its crystals are cubes, as shown in the cut.

GALENA.

White-Lead Ore, or *Carbonate of Lead*, is a whitish mineral, not often very abundant, but very rich in lead, and very valuable as an ore. It makes a valuable pigment.

Phosphate of Lead, or *Pyromorphite*, is a bright-green or brown mineral, sometimes taking an orange-yellow tint. It is not uncommon in lead districts, but not often abundant.

There are at least thirty minerals in which lead is the important and characterising ingredient, but though interesting none of them have much value in the arts.

Ores of Zinc.

Zinc is a metal much used in the arts, and having some analogies with lead, though in many respects exceedingly different from it. Like lead, it is employed as a metal, though never found native. It oxidises on exposure to damp air. It can easily be distilled. Its colour is bluish-white, and it breaks with a very brilliant fracture. In its metallic form it has only been recognised since the thirteenth century, but its ore, calamine, has been used from time immemorial to modify copper, and produce the important mixed metal called Brass. It forms also other alloys, but this is, beyond comparison, the most important. At certain temperatures, or rather within certain limits of temperature, zinc is malleable, but when cold or hot it is brittle, and between 400° and 500° F. becomes so brittle that it can be pounded in a mortar. Zinc is often used as a substitute for lead, as it is much lighter, and to some extent answers the same purpose. It is also extensively used to coat iron with the view of preserving it from rust.

Calamine is the principal ore of zinc. It is a carbonate of the oxide, and occurs as a crystalline mineral of impure white or greenish-white colour, often almost transparent and brittle. When pure it contains 65 per cent. of the oxide of zinc (about 52 per cent. of metal), but it is generally impure. It is found with galena in many lead-mines, and has existed in large quantity in Belgium and on the Rhine, where the supply is now almost exhausted. It is mined in Greece, in China, and in the United States.

Blende is now frequently used as an ore of zinc. It is the sulphide or combination with sulphur, the proportion of metal being 67 per cent. when the ore is pure. It is, however, rarely met with so rich. It is always crystalline, and is of yellow, red-brown, or black colour. It is known to miners as Black Jack. Blende is very usually found with galena, and the ores of zinc often occupy the upper part of lead-veins. Sometimes blende and calamine, but more generally blende and galena, are associated. The crystals of blende differ materially from those of galena, as will be seen by looking at the annexed cut and comparing it with that on p. 265.

BLENDE.

Spartalite is the name of an oxide of zinc, containing 75 per cent of metal. It is red or yellow when impure, but colourless when pure. It is a good ore, but only found hitherto in the State of New Jersey, U.S.

CHAPTER X.

USEFUL MINERALS BEING ORES OF THE LESS IMPORTANT METALS.

THERE are in addition to the metals already described eight others, which are, to some extent, utilised either directly or indirectly; and besides these five others, which have at present no known place in manufacture or the arts. This apparent absence of utility of so many supposed elementary substances which help to make up the earth's crust may arise from either of two causes: either, on the one hand, the missing minerals are not really elements, but combinations, or, on the other hand, the earth being only one member of a system, these elements are present with us without being essential, their main use in creation being elsewhere. Several of the metals that will be alluded to in this chapter possess only a secondary interest; but others are decidedly important, though not as metals. A few words regarding each will be necessary.

Manganese.

This metal can only be obtained in a metallic state by careful chemical processes, and cannot be retained as a metal exposed to the atmosphere, owing to the extreme rapidity with which it decomposes water, taking up the oxygen, and liberating the hydrogen. It is thus hardly known as a metal, and is of no use in that state. It is produced occasionally by the chemist, and is something like cast-iron, brittle, very difficult to melt, and so hard as to scratch steel.

The readiness with which manganese combines with oxygen, and the facility with which it parts from it, are the causes why the oxides are exceedingly useful in the arts. They are employed in bleaching and in glass-making, giving a violet colour to glass.

Pyrolusite, the most common ore of manganese, is an oxide. It is common in most parts of the world, and gives off about 10 per cent. of oxygen at a red heat. It is a very useful mineral *Wad*, another oxide, but with water, is also an important ore, but very impure. It is used

MANGANESE ORE.

in bleaching and in the manufacture of *umber*, an important pigment. *Psilomelane* is a mixture of the oxide of manganese with baryta. It is also abundant. *Manganese Spar* is a silicate, often crystalline. It is the most beautiful of the manganese ores, and is capable of taking a high polish, so that it is used sometimes as an ornamental stone. Its chief use, however, is in colouring glass and glazing pottery. It is common, and is found both in veins and beds.

Manganese appears to form an important ingredient in the iron ore best fitted to produce what is known as Bessemer steel. A kind of pig-iron called *Spiegeleisen* cnotains as much as 10 per cent. of manganese, and is chiefly obtained from the Rhenish provinces. This is mixed in certain proportions with the metal to be converted, and is an essential ingredient.

Nickel.

This metal, though not otherwise found native, is usually present in meteoric iron. Its colour is a brilliant, but greyish-white, sufficiently near silver to form a tolerable substitute. It is malleable and ductile. At low temperatures it is as magnetic as iron, but it loses this property when heated. It is even less fusible than iron, and does not oxidise on exposure.

Nickel is a valuable and useful metal. Alloyed with copper and zinc it becomes *German Silver*, which is very extensively used as a substitute for silver plate either alone or electro-plated. *White Copper* is a similar alloy, and *Packfong* or *Tutenague* another. In these, however, the nickel is only in small proportion. Some of the white metals have long been in use in China.

The only ore of Nickel is called *Copper Nickel* or *Arsenical Nickel*. It contains 44 per cent. of the metal and a still larger percentage of arsenic, but the nickel being the more important metal it properly belongs to this head. While the nickel also is permanent, the arsenic is sometimes replaced by antimony. Cobalt is always, and silver frequently, present in it. The mineral is of a pale copper-red colour and metallic lustre. *Garnierite* is the name of a new ore of nickel found abundantly in New Caledonia, and used in the manufacture of nickel bronze, an alloy half nickel half bronze well adapted for use in many manufactures.

There is another combination of nickel sometimes met with, and found in considerable quantities in Scotland. It is a sulphide of iron and nickel, forming a nickeliferous iron pyrites containing 14 per cent. of nickel.

Cobalt.

Cobalt, like nickel, is only found in the metallic state in meteoric iron. It is of reddish-steel-grey colour, brittle, and rather soft, but capable of taking a high polish. Like nickel it is less fusible than iron, and changes less on exposure, but as a metal it has never entered into use in the arts. The use of cobalt is entirely confined to the oxides, which yield exceedingly valuable pigments and a colouring material for glass and porcelain. The exquisite blue derived from cobalt cannot be matched, and is very permanent. The blues of cobalt are called *smalt* or *zaffre* in the form in which they are brought into the market, and they are obtained partly from the Arsenide (*Arsenical* or *Tin white Cobalt*, containing sometimes 23 per cent. cobalt and the rest chiefly arsenic), partly from *Cobalt bloom*, and partly from *Red cobalt ochre*, these latter being earthy arseniates of cobalt. All these ores are comparatively rare and valuable, but except the latter they give no indication of the colour derived from them. *Zaffre* is an impure oxide of an intense blue colour, and *smalt* a pounded blue glass, composed of three parts of a mixture of zaffre with sand and potash. So deep is the colour, that one grain of zaffre will give a full blue to 240 grains of glass.

The most abundant ore of cobalt is that first mentioned. It is obtained chiefly from Saxony, Bohemia, and Sweden. *Cobalt glance* is found at Siegen, in Rhenish Prussia. The earthy ores come from Saxony, and are probably modifications of the arsenide. As a rule, ores of cobalt are so much mixed up with those of nickel that the preparations of the two metals are obtained from them together by a mixed process.

Antimony.

Antimony is a metal of some importance, but is so seldom found in the metallic state that its appearance in that form may be neglected. It is also so common combined with sulphur that the sulphide is the recognised form. It is a silver-white metal with brilliant lustre, about the hardness of gold, compact and brittle. It does not oxidise on exposure at ordinary temperature, but a little below red heat it fuses and burns brilliantly. Its use as a metal is confined to the manufacture of type-metal, hard pewter, and Britannia metal, being mixed with lead, tin, and copper, to produce these alloys. It also forms a hard alloy with iron. It renders hard and brittle any metal with which it is combined. The oxides are used in medicine and in colouring paste gems.

Grey Antimony, the sulphide, is the only ore. It is an important mineral, generally in bunches of long prismatic crystals, easily recognised, and very metallic in appearance. It melts in the flame of a candle. It is not uncommon in the British islands, and is found in various places in France and Hungary.

Very large quantities exist in Borneo. From it is obtained the antimony of commerce.

This ore was called *Stibium* by the ancients, and is now called *Stibnite* in mineralogical works. It was anciently used for darkening the upper and under sides of the eyelids for the purpose of increasing the apparent size of the eye, being supposed not only to impart additional beauty and brilliancy to the eye and make it appear larger, but to be also beneficial to the sight. The practice of staining the eyelids is not only a very ancient practice in the East, but is common among the women of Syria at the present day. It is often alluded to in the Bible, as we read in the history of Ahab that "when Jehu was come to Jezreel, Jezebel heard of it, and she painted her face [literally "put her eyes in painting"], and tired her head, and looked out at a window" (2 Kings ix. 30). And again in the prophecy of Ezekiel in reproving the adulteries of Aholah and Aholibah we read, "For whom thou didst wash thyself, paintedst thine eyes, and deckedst thyself with ornaments" (Ezekiel xxiii. 40). In this passage the Hebrew expression as rendered in the Septuagint is literally "thou didst paint thine eyes with stibium," and in the Vulgate the words also are *circumlinisti stibio oculos tuos*, "thou didst paint round thine eyes with stibium." In the prophecy of Jeremiah (iv. 30) there is an expression, not very clear in our translation, but which alludes to a curious result that happens occasionally by the careless use of antimonial powder. The expression is "though thou rentest thy face with painting, in vain shalt thou make thyself fair." While using the powder of antimony, some of it frequently enters the eye itself and produces great irritation.

Antimony was also used as a hair-dye by the ancients, and was employed to colour the eyebrows. The mode of preparing it is described by Dioscorides, and consisted in enclosing the powder in a lump of dough and burning it on the coals till it was reduced to a cinder. It was then extinguished with milk and wine, and again placed on the coals and the blast repeated till ignition took place, after which the heat was allowed to subside lest the mass should be reduced to the metallic state.

Chromium.

Chromium is a hard, brittle, and rather heavy metal, resembling iron, standing exposure to the air without injury, exceedingly infusible, and having high metallic lustre. It has never been utilized as a metal in any way, and though found in meteoric iron, with nickel and other elements, has not been met with native in a detached form. It is never used as an alloy.

Like some other metals, the preparations of chromium are valuable as pigments, and for this purpose the oxides and certain salts are used. The only ore is called *Chrome Iron* or *Chromite*. It is found often in serpentines and other rocks of which magnesia is a component part, and it gives the green colour to that mineral. Though nowhere found in large quantity, it is met with in various places in Germany, in Norway,

Hungary, Scotland, and the Ural Mountains, and also in the United States. In the various ores the oxide of chrome is present to an extent varying from 35 to 60 per cent. The colour of the oxide is green, and it is generally crystallized in octahedra. The ore, though called chromate of iron, is a compound of the oxides of chromium and iron. It is sometimes magnetic, and has a resinous appearance, and a brownish-black colour. From it is produced chromate of potash, from which salt the various useful preparations are derived.

Chromium salts are valuable for the yellow and green colours obtained from them.

Uranium.

Uranium is a much more rare metal, in all its forms, than chromium, and though white, malleable, and not oxidised by air or water, is incapable of being preserved under ordinary conditions in the air, as it is very combustible when exposed to heat, and unites with great violence with sulphur or chlorine. The ores are oxides, the principal one being called *Pitchblende*, a brownish or blackish mineral, velvety in appearance, and found in lead and silver veins in Saxony, and in tin veins in Cornwall.

The use of Uranium oxide is confined to enamel-painting and glass-staining, and the colours obtained are either black or delicate yellow. Combined with copper it yields a blue colour, sometimes used in paper-making.

Tungsten.

This metal is not found native, but occurs in the form of *Wolfram*, a mineral frequently associated with tin. In this form it is considered a great deal too common in Cornwall, where the mechanical treatment of tin-ore is rendered much more difficult than it would otherwise be in consequence of its presence. The reason of this is the small difference in specific gravity between the two minerals. By chemical treatment, calcining the mixed ores of tin and tungsten with soda-ash, the tungsten is however got rid of by washing as a soluble salt of soda, after which other impurities are easily separated.

Tungsten is remarkable as being one of the heaviest metals known, being almost equal to gold in this respect. It is also excessively difficult of fusion. It is hard and brittle, and has not been obtained in large quantity. Though of no use as a separate metal, it produces some remarkable alloys, especially with iron, and it improves the mixed metal called pack-fong, which is a mixture of copper and tin, better known as Britannia metal.

Tungstate of soda is used in dyeing, chiefly as a mordant, by which is meant a preparation whose object is to preserve certain parts of a material from the influence of a dye into which the whole is dipped. It has also been found to prevent light cotton and linen manufactures from burning, and a patent was obtained for the use of it in rendering ladies' dresses non-inflammable. It does not appear to have been much used for this purpose, as we still hear from time to time of accidents by fire, owing to carelessness in the use of such material.

Molybdenum.

The relations of Molybdenum to Tungsten are very close, the two replacing each other in some of the minerals containing them. Molybdenum occurs generally with Wolfram, but it is also combined with lead. It is sometimes found as a sulphide, in which form it greatly resembles graphite. It is of no use in the arts. The name is derived from the resemblance it has to lead.

Cadmium.

Cadmium is a metal very generally found with zinc-ores, and having some relations with that and other metals. It resembles tin in colour and lustre, and takes a fine polish, but it is soft, and, like lead, it soils anything rubbed on it. Like tin it makes a peculiar creaking sound when bent. It is heavy and very fusible.

The only ore of cadmium is the sulphide, and is called *Greenockite*. This is a rare mineral, but so many of the ores of zinc contain cadmium associated with them that the metal itself, if of any value, need hardly be rare. Its salts are used to some extent in medicine, and as a yellow pigment. It mixes readily with several metals, such as copper, platinum, mercury, lead, and bismuth. Some of the alloys are exceedingly fusible, but except as a material for stopping teeth, and as forming an ingredient in some fusible metals, none of them can be said to be utilized.

Besides these metals, most of them of small or no importance in the metallic state, and others only interesting as slightly modifying other metals, there are several that are known, but of yet smaller importance in the arts. The most that can be said of them at present is that they are the bases of rare minerals. Their names alone are of little interest, but a few of their compounds enter into preparations more or less familiar. Thus, *Vanadium* is used as vanadate of ammonia in making a valuable writing-ink. It occurs with lead, in Cheshire, as a vanadate of lead. *Tellurium* is associated with gold and bismuth, but only in modifying those metals, and at present has no value. *Tantalum*, *Niobium*, *Pelopium*, and *Yttrium* are metals equally little known, and even less interesting, except to the chemist.

Artificial Production of Diamonds.

While the sheets of this work were actually in the press satisfactory proof was obtained and made known that carbon can be crystallised artificially and true crystals of diamond manufactured.

This result, obtained by very ingenious methods with great difficulty, and after experiments by various chemists, is not one that need alarm the dealers in, or possessors of, these beautiful gems. Where Nature has so sparingly produced the perfect crystalline condition of a very common elementary substance, it is unlikely that there can be any easy or inexpensive way of bringing about the same result. The discovery, however, is as important as it is interesting.

February 24, 1880. D. T. A.

INDEX.

	PAGE
Adularia	135
Agalmatolite	139
Agate	110
Alabaster	150
Alabastritis	148
Alexandrite	75
Allophane	216
Almandine	89
Almandine ruby	82
Alumstone	210
Amalgam, native	234
Amazon stone	134
Amber	161
Amethyst	39, 61, 97
Amianthus	209
Anthrax	89
Antimony ores	272
Apatite	205
Aplome	88
Aquamarine	72, 73
Arsenic, native	242
Arsenical cobalt	271
Arsenical nickel	270
Arsenical silver	231
Asbestos	208
Asparagus stone	205
Asteria	66
Augite	208
Auro-tellurite	230
Aventurine	101, 134
Azure stone	129
Azurite	262
Balas ruby	61, 82
Barytes	203
Basanite	117
Bell metal	239
Beryl	39, 72
Bismuth, native	241
Black band	257
Black Jack	267
Black lead	185
Black oxide of copper	262
Black silver	232
Blende	267
Blood stone	116
Blue John	154

	PAGE
Blue topaz	81
Blue vitriol	262
Bog iron ore	255
Bohemian garnets	87
Boort	59
Boracite	202
Borax	201
Borea	116
Bort	59
Brass	239
Brazilian sapphire	81
Brocatello	145
Brilliants	57
Bristol diamond	96
Brittle silver ore	232
Bronze	239
Brown hæmatite	255
Brown ochre	255
Cacholong	121
Cadmium	277
Cairngorm	100
Calamine	266
Calcite	140
Callais	127
Cannel coal	161
Carbuncle	65, 89
Carnelian	39, 105
Carrara marble	142
Cascalho	51
Cassiterite	263
Cat's-eye	114
Celestine	203
Ceraunite	66
Chalcedonic quartz	102
Chalcedony	39, 102
Chatoyant sapphires	61
Chessylite	131
Chrome iron	274
Chromite	274
Chrysoberyl	74
Chrysocolla	132
Chrysolite	39, 124
Chrysoprase	39, 104
Cingalese garnets	87
Cinnabar	234
Cinnamon stone	77, 87

	PAGE		PAGE
Cipolino	145	Golden beryl	74
Citrine	100	Golden opal	120
Clay ironstones	257	Grammatias	116
Cobalt bloom	271	Graphite	185
Cobalt glance	272	Greenockite	277
Copper	238	Grey antimony	272
Copper glance	260	Grey copper ore	261
Copper nickel	270	Grossular	87
Copper pyrites	258, 260	Gypsum	152
Coprolites	205	HÆMATITE	252
Coral	179	Harlequin opal	120
Cornalline	106	Heavy spar	203
Cornish diamonds	96	Heliotrope	116
Corundum	210	Hornblende	208
Cotham marble	114, 145	Horn silver	232
Crystal	92	Hyacinth	39, 63, 76
Cupid's net	102	Hyalite	121
Cupid's pencils	102	Hydrophane	121
Cyanite	211	ICELAND spar	140
Cyanos	130	Idocrase	126
Cymophane	75	Imitative gems	90
DOLOMITE	204	Iridium	237
EGYPTIAN onyx	145	Iridosmine	238
Electron	162	Iris	97
Electrum	230	Irish diamonds	96
Emerald	39, 68	Irite	238
Emery	210	Iron, meteoric	244
Enhydros	95	Iron pyrites	258
Euclase	75	Itacolumite	52
FAHLERZ	261	JACINTH	39, 76
False topaz	100	Jacotinga	230
Felspar	216	Jade	137
Fibrous red iron ore	255	Jargoon	76
Fire clay	212	Jasper	39, 114
Fire garnet	89	Jaspery quartz	114
Fire marble	145	Jet	156
Fire opal	120	KAOLIN	211
Flints	118	Kish	189
Fluor spar	154	Kyanite	211
Fortification agate	112	LABRADOR felspar	133
Fuller's earth	211, 215	Labradorite	133
GALENA	265	Landscape marble	145
Garnet	85	Landscape stone	114
German silver	239	Lapis Basanitis	117
German topaz	100	Lapis Lazuli	129
Giallo antico	145	Lead glance	265
Girasol	120	Lead ores	264
Girasol sapphires	61	Lignite	161
Glucina	69	Lime	204
Gœthite	136	Limestone	205
Gold	221	Limonite	255

INDEX.

	PAGE
Litharge	264
Lodestone	250
Love's meshes	102
Lumachelle	145
Lychnis	64, 79
Lydian stone	117
Lyncurium	78, 164
MAGNESITE	207
Magnetic iron ore	250
Malachite	130, 262
Manganese spar	269
Marble	141
Marcasite	258
Masticot	264
Meerschaum	207
Melanite	88
Mercury	233
Meteoric iron	244
Mexican onyx	145
Micaceous iron ore	254
Minium	264
Mispickel	258
Mocha stone	113
Molochites	132
Molybdenum	277
Moonstone	135
Morion	101
Moss agate	113
Mother-of-pearl	169
Müller's glass	121
Mundic	258
NACRE	169
Native amalgam	234
Native arsenic	242
Native bismuth	241
Native copper	238
Native gold	221
Native Prussian blue	258
Native silver	230
Natron	199
Nephrite	137
Nero antico	144
Nickel ores	270
Nicolo	109
Niobium	278
Nitrate of soda	199
Nitre	200
Noble opal	118
Noble serpentine	136
OBSIDIAN	217
Occidental turquoise	129
Oculus mundi	121
Odontolite	128
Oligist	254
Olivine	125
Onicolo	109
Onyx	107
Opal	118
Opalescent sapphires	61
Ophiolite	137
Ophite	137
Oriental alabaster	148
Oriental amethyst	61
Oriental chrysolite	125
Oriental opal	118
Orpiment	242
Osmiridium	238
Osmium	237
PACKFONG	270
Palladium	237
Parian marble	142
Pearl	168
Pelopium	278
Pentelic marble	142
Peridote	125
Phenacite	75
Phosphate of lead	265
Phosphorite	205
Pingos d'agua	81
Pipe clay	212
Pitch blende	275
Plasma	104
Plastic clay	212
Platiniridium	238
Platinum	235
Pleonaste	83
Plumbago	185
Prase	104
Precious garnet	89
Precious opal	118
Precious serpentine	136
Psilomelane	269
Pumice-stone	217
Pyrolusite	269
Pyromorphite	265
Pyrope	89
Pyrophane	122
QUARTZ	92
REALGAR	242
Red cobalt ochre	271
Red copper ore	261
Red hæmatite	255

	PAGE		PAGE
Red lead	264	Steatite	139
Red ochre	255	Steel	247
Reddle	255	Stibium	273
Rhodium	237	Stibnite	273
Ribbon agate	112	Stream tin	264
Rock crystal	92	Strontian	204
Rock salt	192	Succinum	164
Rose diamond	57	Sulphur	189
Rose quartz	99	Sunstone	135
Rosso antico	144	Syrian garnets	87
Rubellite	84	TABASHEER	123
Rubicelle	82	Tantalium	278
Ruby	62	Tellurated silver	231
Ruby silver	232	Tellurium	278
Ruin marble	145	Tenorite	262
Ruthenium	238	Thunderstone	67
SALTPETRE	200	Tin ores	262
Sapphire	130	Tin-white cobalt	271
Sard	62, 107	Topaz	39, 79
Sardonyx	39, 109	Topazolite	88
Sassoline	202	Tourmaline	79, 84
Saxon topaz	80	Tungsten	276
Schorl	85	Turkey stone	215
Selenite	135, 152	Turquoise	126
Selenium	190	Tutenague	270
Semi-opal	122	UMBER	256, 269
Serpentine	136	Uranium	275
Shell marble	145	VANADIUM	278
Siderite	257	Venus' hair stone	102
Siena marble	145	Venus' pencils	102
Silver fahlerz	261	Verde antico	144
Silver glance	232	Vermilion	234
Silver, native	230	Vitreous copper ore	260
Slate	215	Vitreous silver	232
Smaragdus	67, 70	Vivianite	258
Smoky quartz	100	WAD	269
Soapstone	139	Whetstone	215
Sparry iron ore	256	White arsenic	242
Spartalite	267	White copper	270
Spathic iron	256	White iron pyrites	258
Specular iron ore	254	White lead ore	265
Spessartine	88	White topaz	81
Sphæro siderite	257	Witherite	203
Spiegeleisen	269	Wolfram	276
Spinelle	82	Wood tin	263
Spinelle ruby	61, 82	YELLOW jacinth	78
Stahlstein	257	Yellow ochre	255
Stalactites	147	Yttrium	278
Stalagmites	147	ZINC ores	266
Star stone	66	Zircon	76

PUBLICATIONS

OF THE

Society for Promoting Christian Knowledge.

THE ROMANCE OF SCIENCE.

Small post 8vo, Cloth boards.

Coal, and what we get from it. Expanded from the Notes of a Lecture delivered at the London Institution. By Professor RAPHAEL MELDOLA, F.R.S., F.I.C. With several Illustrations. 2s. 6d.

Colour Measurement and Mixture. By Sir W. DE W. ABNEY, K.C.B., R.E., F.R.S. Numerous Illustrations. 2s. 6d.

The Making of Flowers. By the Rev. Professor GEORGE HENSLOW, M.A., F.L.S., F.G.S. Several Illustrations. 2s. 6d.

The Birth and Growth of Worlds. A Lecture by Professor A. H. GREEN, M.A., F.R.S. 1s.

Soap-Bubbles, and the Forces which Mould Them. A course of Lectures by C. V. BOYS, A.R.S.M., F.R.S. With numerous diagrams. 2s. 6d.

Spinning Tops. By Professor J. PERRY, M.E., D.Sc., F.R.S. With numerous diagrams. 2s. 6d.

Our Secret Friends and Foes. By PERCY FARADAY FRANKLAND, Ph.D., B.Sc. (Lond.), F.R.S. Fourth Edition, revised and enlarged. 3s.

Diseases of Plants. By Professor MARSHALL WARD. With numerous Illustrations. 2s. 6d.

Sounding the Ocean of Air. Being six Lectures delivered before the Lowell Institute of Boston, in December 1898, by A. LAWRENCE ROTCH, S.B., A.M. With numerous Illustrations. 2s. 6d.

The Story of a Tinder-Box. By the late MEYMOTT TIDY, M.B., M.S., F.C.S. Numerous Illustrations. 2s.

Time and Tide. A Romance of the Moon. By Sir ROBERT S. BALL, LL.D., Royal Astronomer of Ireland. With Illustrations. 2s. 6d.

The Splash of a Drop. By Professor A. M. WORTHINGTON, F.R.S. 1s. 6d.

NATURAL HISTORY RAMBLES.

Fcap. 8vo., with numerous Woodcuts, Cloth boards, 2s. 6d. each.

IN SEARCH OF MINERALS.
By the late D. T. ANSTEAD, M.A., F.R.S.

LANE AND FIELD.
By the late Rev. J. G. WOOD, M.A., Author of "Man and his Handiwork," &c.

MOUNTAIN AND MOOR.
By J. E. TAYLOR, F.L.S., F.G.S., Editor of "Science-Gossip."

PONDS AND DITCHES.
By M. C. COOKE, M.A., LL.D.

THE SEA-SHORE.
By Professor P. MARTIN DUNCAN, M.B. (London), F.R.S.

THE WOODLANDS.
By M. C. COOKE, M.A., LL.D., Author of "Freaks and Marvels of Plant Life," &c.

UNDERGROUND.
By J. E. TAYLOR, F.L.S., F.G.S.

HEROES OF SCIENCE.
Crown 8vo. Cloth boards, 4s. each.

ASTRONOMERS. By E. J. C. MORTON, B.A. With numerous diagrams.
BOTANISTS, ZOOLOGISTS, AND GEOLOGISTS. By Professor P. MARTIN DUNCAN, F.R.S., &c.
CHEMISTS. By M. M. PATTISON MUIR, Esq., F.R.S.E. With several diagrams.
MECHANICIANS. By T. C. LEWIS, M.A.
PHYSICISTS. By W. GARNETT, Esq., M.A.

MAPS.

MOUNTED ON CANVAS AND ROLLER, VARNISHED.

		s.	d.
EASTERN HEMISPHERE ....4 ft. 10 in. by 4 ft. 2 in.		13	0
WESTERN HEMISPHERE	ditto.	13	0
EUROPE	ditto.	13	0
ASIA. Scale, 138 miles to an inch......	ditto.	13	0
AFRICA	ditto.	13	0
NORTH AMERICA. Scale, 97 m. to in.	ditto.	13	0
SOUTH AMERICA. Scale, ditto.	ditto.	13	0
AUSTRALASIA..	ditto.	13	0
AUSTRALASIA (Diocesan Map).	ditto.	14	0
INDIA. Scale, 40 m. to in. 50 in. by 58 in.......................		13	0
AUSTRALIA	ditto.	9	0
IRELAND. Scale, 8 m. to in., 2 ft. 10 in. by 3 ft. 6 in.		9	0
SCOTLAND. Scale, ditto...............	ditto.	9	0
GREAT BRITAIN AND IRELAND,			
The United Kingdom of .... 6 ft. 3 in. by 7 ft. 4 in.		42	0
ENGLAND AND WALES (Photo-Relievo) 4 ft. 8 in. by 3 ft. 10 in.		13	0
ENGLAND AND WALES (Diocesan Map) 4 ft. 2 in. by 4 ft. 10 in.		16	0
BRITISH ISLES .... 58 in. by 50 in.		13	0
HOLY LAND .... 4 ft. 2 in. by 4 ft. 10 in.		13	0
SINAI (The Peninsula of), the NEGEB, and LOWER EGYPT. To illustrate the History of the Patriarchs and the Exodus............................2 ft. 10 in. by 3 ft. 6 in.		9	0
PLACES mentioned in the ACTS and the EPISTLES. Scale, 57 miles to an inch .... 3 ft. 6 in. by 2 ft. 10 in.		9	0
Photo-Relievo Maps, on Sheets, 19 in. by 14 in. :—			
ENGLAND AND WALES. SCOTLAND. EUROPE.			
Names of places and rivers left to be filled in by scholars.. each		0	6
With rivers and names of places........................ ,,		0	9
With names of places, and with county and country divisions in colours .... ,,		1	0
AFRICA. With rivers and names of places, etc............ ,,		0	9
ASIA AND NORTH AMERICA.			
Names of places and rivers left to be filled in .... ,,		0	6
With rivers and names of places......................... ,,		0	9
NORTH LONDON. With names of places, etc............. ,,		0	6
SOUTH LONDON. With names of places, etc............... ,,		0	6
PHOTO-RELIEVO WALL MAP. ENGLAND AND WALES. 56 in. by 46 in. *on canvas roller and varnished,* *plain* 12s., *coloured* ,,		13	0

ATLASES.

                                                                         s. d.

HANDY GENERAL ATLAS OF THE WORLD (The). A Comprehensive series of Maps illustrating General and Commercial Geography. With Complete Index...*Half morocco* 42 0

BIBLE ATLAS (The). Sixth Edition, Revised, by Sir CHARLES WILSON, K.C.B., K.C.M.G., F.R.S. Royal 4to. *Cloth boards* 10 6

A MODERN ATLAS: containing 30 Maps, with Indexes, etc..*Cloth boards* 12 0

HANDY REFERENCE ATLAS OF THE WORLD, with Index and Geographical Statistics......*Cloth boards* 7 6

STAR ATLAS (The). Translated and adapted from the German by the Rev. E. MCCLURE. With 18 Charts. *Cloth* 7 6

WORLD (The), an **ATLAS**, containing 34 Coloured Maps and Complete Index. Folded 8vo..............*Cloth gilt* 5 0

HANDY ATLAS OF THE COUNTIES OF ENGLAND. Forty-three Coloured Maps and Index...........*Cloth* 5 0

CENTURY ATLAS OF THE WORLD. A Series of 66 Maps. With General Index and Geographical Statistics. Edited by J. G. BARTHOLOMEW, F.R.G.S., F.R.S.E. New Edition, brought up to date. 4to..*Cloth* 3 6

MINIATURE ATLAS AND GAZETTEER OF THE WORLD..*Cloth* 2 6

POCKET ATLAS OF THE WORLD (The). With Complete Index, etc. ....*Cloth* 2 6

BRITISH COLONIAL POCKET ATLAS (The). Fifty-six Maps of the Colonies, and Index ....*Cloth boards* 2 6
                                  Paste grain roan 3 6

PHYSICAL ATLAS FOR BEGINNERS, containing 12 Coloured Maps........ ....*Paper cover* 1 0

SIXPENNY BIBLE ATLAS (The), containing 16 Coloured Maps... ....*Paper wrapper* 0 6

SHILLING QUARTO ATLAS (The), containing 24 Coloured Maps................................*Paper wrapper* 1 0

BRITISH COLONIES (Atlas of the), containing 16 Coloured Maps.............................*Paper cover* 0 6

PENNY ATLAS (The), containing 13 Maps.......Small 4to. 0 1

MANUALS OF HEALTH.

Fcap. 8vo, 128 pp., Limp Cloth, price 1s. each.

HEALTH AND OCCUPATION. By the late Sir B. W. RICHARDSON, F.R.S., M.D.

HABITATION IN RELATION TO HEALTH (The). By F. S. B. CHAUMONT, M.D., F.R.S.

NOTES ON THE VENTILATION AND WARMING OF HOUSES, CHURCHES, SCHOOLS, AND OTHER BUILDINGS. By the late ERNEST H. JACOB, M.A., M.D. (OXON.).

ON PERSONAL CARE OF HEALTH. By the late E. A. PARKES, M.D., F.R.S.

AIR, WATER, AND DISINFECTANTS. By C. H. AIKMAN, M.A., D.Sc., F.R.S.E.

MANUALS OF ELEMENTARY SCIENCE.

Foolscap 8vo, 128 pp. with Illustrations, Limp Cloth, 1s. each.

PHYSIOLOGY. By Professor A. MACALISTER, LL.D., M.D., F.R.S., F.S.A.

GEOLOGY. By the Rev. T. G. BONNEY, M.A., F.G.S. New and Revised Edition.

ASTRONOMY. By W. H. CHRISTIE, M.A., F.R.S.

BOTANY. By the late Professor ROBERT BENTLEY.

ZOOLOGY. By ALFRED NEWTON, M.A., F.R.S., Professor of Zoology in the University of Cambridge. New Revised Edition.

MATTER AND MOTION. By the late J. CLERK MAXWELL, M.A., Trinity College, Cambridge.

SPECTROSCOPE (THE), AND ITS WORK. By the late RICHARD A. PROCTOR.

CRYSTALLOGRAPHY. By HENRY PALIN GURNEY, M.A., Clare College, Cambridge.

ELECTRICITY. By the late Prof. FLEEMING JENKIN.

MISCELLANEOUS PUBLICATIONS.

                                                                                                      s. d.

Among the Birds. By FLORENCE ANNA FULCHER.
Large Crown 8vo..*Cloth boards* 3 6

Animal Creation (The). A popular Introduction to Zoology. By the late THOMAS RYMER JONES, F.R.S. With 488 Woodcuts. Post 8vo................*Cloth boards* 7 6

Birds' Nests and Eggs. With 22 coloured plates of Eggs. Square 16mo........................*Cloth boards* 3 0

British Birds in their Haunts. By the late Rev. C. A. JOHNS, B.A., F.L.S. With 190 engravings by Wolf and Whymper. Post 8vo................*Cloth boards* 5 0

Evenings at the Microscope; or, Researches among the Minuter Organs and Forms of Animal Life. By the late PHILIP H. GOSSE, F.R.S. A New Edition, revised by Professor F. JEFFREY BELL. With numerous Illustrations. Crown 8vo........................... *Cloth boards* 5 0

Fern Portfolio (The). By FRANCIS G. HEATH, Author of "Where to find Ferns," &c. With 15 plates, elaborately drawn life-size, exquisitely coloured from Nature, and accompanied with descriptive text.
                                                           Cloth boards 8 0

Fishes, Natural History of British; their Structure, Economic Uses, and Capture by Net and Rod. By the late FRANK BUCKLAND. With numerous Illustrations. Crown 8vo..*Cloth boards* 5 0

Flowers of the Field. By the late Rev. C. A. JOHNS, B.A., F.L.S. (29th edition.) Entirely rewritten and revised by Professor G. S. BOULGER, F.L.S., F.G.S., Professor of Botany in the City of London College. Post 8vo.....................................*Cloth boards* 7 6

SOCIETY FOR PROMOTING CHRISTIAN KNOWLEDGE. 7

                  s. d.

Forest Trees (The) of Great Britain. By the late Rev. C. A. JOHNS, B.A., F.L.S. With 150 woodcuts. Post 8vo..*Cloth boards* 5 0

Freaks and Marvels of Plant Life; or, Curiosities of Vegetation. By M. C. COOKE, M.A., LL.D. With numerous illustrations. Post 8vo...............*Cloth boards* 6 0

Man and his Handiwork. By the late Rev. J. G. WOOD, Author of "Lane and Field," &c. With about 500 illustrations. Large Post 8vo. ....*Cloth boards* 7 6

Natural History of the Bible (The). By H. B. TRISTRAM, D.D., LL.D., F.R.S. With numerous illustrations. Crown 8vo.................................*Cloth boards* 5 0

Nature and her Servants; or, Sketches of the Animal Kingdom. By the Rev. THEODORE WOOD. With numerous woodcuts. Large Post 8vo. *Cloth boards* 4 0

Ocean (The). By the late PHILIP H. GOSSE, F.R.S., Author of "Evenings at the Microscope." With 51 illustrations and woodcuts. Post 8vo...... .*Cloth boards* 3 0

Our Bird Allies. By the Rev. THEODORE WOOD. With numerous illustrations. Fcap. 8vo...*Cloth boards* 2 6

Our Insect Allies. By the Rev. THEODORE WOOD. With numerous illustrations. Fcap. 8vo. *Cloth boards* 2 6

Our Insect Enemies. By the Rev. THEODORE WOOD. With numerous illustrations. Fcap. 8vo. *Cloth boards* 2 6

Our Island Continent. A Naturalist's Holiday in Australia. By J. E. TAYLOR, F.L.S., F.G.S. With Map. Fcap. 8vo. ....*Cloth boards* 2 6

		s.	d.

Our Native Songsters. By ANNE PRATT, Author of "Wild Flowers." With 72 coloured plates. 16mo. .... *Cloth boards* 4 0

Romance of Low Life amongst Plants. Facts and Phenomena of Cryptogamic Vegetation. By M. C. COOKE, M.A., LL.D., A.L.S. With numerous woodcuts. Large post 8vo. .... *Cloth boards* 4 0

Selborne (The Natural History of). By the REV. GILBERT WHITE. With Frontispiece, Map, and 50 woodcuts. Post 8vo. .... *Cloth boards* 2 6

Toilers in the Sea. By M. C. COOKE, M.A., LL.D. Post 8vo. With numerous illustrations. ..... *Cloth boards* 5 0

Vegetable Wasps and Plant Worms. By M. C. COOKE, M.A. Illustrated. Post 8vo. .... *Cloth boards* 5 0

Wayside Sketches. By F. EDWARD HULME, F.L.S. With numerous illustrations. Crown 8vo. *Cloth boards* 4 0

Where to find Ferns. By FRANCIS G. HEATH, Author of "The Fern Portfolio," &c. With numerous illustrations. Fcap. 8vo. .... *Cloth boards* 1 6

Wild Flowers. By ANNE PRATT, Author of "Our Native Songsters," &c. With 192 coloured plates. In two volumes. 16mo. .... *Cloth boards* 8 0

LONDON:
NORTHUMBERLAND AVENUE, CHARING CROSS, W.C.;
43, QUEEN VICTORIA STREET, E.C.

www.ingramcontent.com/pod-product-compliance
Lightning Source LLC
Chambersburg PA
CBHW032049230426
43672CB00009B/1538